INTERNATIONAL CODE OF SIGNALS

As adopted by the Fourth Assembly of the Inter-Governmental Maritime
Consultative Organization in 1965

FOR VISUAL, SOUND, AND RADIO COMMUNICATIONS

UNITED STATES EDITION

1969 Edition

Revised 2003

Reprinted by Starpath Publications

PREFACE

Pub 102, the 1969 edition of the International Code of Signals, became effective on 1 April 1969, and at that time superseded H.O. Pubs. 103 and 104, International Code of Signals, Volumes I and II. All signals are contained in a single volume suitable for all methods of communication.

The First International Code was drafted in 1855 by a Committee set up by the British Board of Trade. It contained 70,000 signals using eighteen flags and was published by the British Board of Trade in 1857 in two parts; the first containing universal and international signals and the second British signals only. The book was adopted by most seafaring nations.

This early edition was revised by a Committee set up in 1887 by the British Board of Trade. The Committee's proposals were discussed by the principal maritime powers and at the International Conference in Washington in 1889. As a result, many changes were made. The Code was completed in 1897 and was distributed to all maritime powers. That edition of the International Code of Signals, however, did not stand the test of World War I.

The International Radiotelegraph Conference at Washington in 1927 considered proposals for a new revision of the Code and decided that it should be prepared in seven languages, namely in English, French, Italian, German, Japanese, Spanish and in one Scandinavian language which was chosen by the Scandinavian Governments to be the Norwegian language. The new edition was completed in 1930 and was adopted by the International Radiotelegraph Conference held in Madrid in 1932. The new Code was compiled in two volumes, one for use by visual signaling and the other by radiotelegraphy. Words and phrases applicable to aircraft were introduced in Volume II together with a complete Medical Section and a Code for accelerating the granting of pratique. The Medical Section and the pratique signals were prepared with the assistance and by the advice of the Office International d'Hygiene Publique. The Code, particularly Volume II, was primarily intended for use by ships and aircraft and, via coastal radio stations, between ships or aircraft and authorities ashore. A certain number of signals were inserted for communications with shipowners, agents, repair yards, etc. The same Conference (Madrid, 1932) established a Standing Committee to review the Code, if and when necessary, to give guidance on questions of use and procedure, and to consider proposals for modifications. Secretarial duties were undertaken by the Government of the United Kingdom. The Standing Committee met only once in 1933 and introduced certain additions and amendments.

The Administrative Radio Conference of the International Telecommunication Union suggested in 1947 that the International Code of Signals should fall within the competence of the Inter-Governmental Maritime Consultative Organization (IMCO). In January 1959, the First Assembly of IMCO decided that the Organization should assume all the functions then being performed by the Standing Committee of the International Code of Signals. The Second Assembly in 1961 endorsed plans for a comprehensive review of the International Code of Signals intended to meet the present day requirements of mariners. A Subcommittee of the Maritime Safety Committee of the Organization was established to revise the Code, to prepare it in nine languages, namely the original seven (English, French, Italian, German, Japanese, Spanish, and Norwegian) together with Russian and Greek, and to consider proposals for a new radiotelephone Code and its relation to the International Code of Signals. The Subcommittee consisted of representatives of the following countries: Argentina, Germany, France, Greece, Italy, Japan, Norway, Russian Federation, United Kingdom, and the United States of America. The following international governmental and nongovernmental organizations contributed to, and assisted in, the preparation of the revised Code: the International Atomic Energy Agency, the International Civil Aviation Organization, the International Labor Organization, the International Telecommunication Union, the World Meteorological Organization, the World Health Organization, the International Chamber of Shipping, the International Confederation of Free Trade Unions, and the International Radio Maritime Committee.

The Subcommittee completed the revision of the Code in 1964, taking into account Recommendation 42 of the 1960 Conference on Safety of Life at Sea and Recommendation 22 of the Administrative Radio Conference, Geneva 1959. The Code was adopted by the Fourth Assembly of IMCO in 1965.

The revised Code is intended to cater primarily for situations related essentially to safety of navigation and persons, especially when language difficulties arise. It is suitable for transmission by all means of communication, including radiotelephony and radiotelegraphy, thus obviating the necessity for a separate radiotelephone Code and dispensing with Volume II for Radiotelegraphy. The revised Code embodies the principle that each signal has a complete meaning. It thus leaves out the vocabulary method which was part of the old Code. The Geographical Section, not being considered essential, was omitted. By these means it was possible to reduce considerably the volume of the Code and achieve simplicity.

Changes and corrections for this product will appear in the NIMA weekly Notice to Mariners and must be applied to keep it current. Users should refer information and comments to: MARITIME SAFETY INFORMATION DIVISION, ST D 44, NATIONAL IMAGERY AND MAPPING AGENCY, 4600 SANGAMORE ROAD, BETHESDA MD 20816-5003.

CONTENTS

CHAPTER **1**

CHAPTER 1
SIGNALING INSTRUCTIONS

CHAPTER 1

SECTION 1: EXPLANATION AND GENERAL REMARKS

1. The purpose of the International Code of Signals is to provide ways and means of communication in situations related essentially to safety of navigation and persons, especially when language difficulties arise. In the preparation of the Code, account was taken of the fact that wide application of radiotelephony and radiotelegraphy can provide simple and effective means of communication in plain language whenever language difficulties do not exist.

2. The signals used consist of:

(a) Single-letter signals allocated to significations which are very urgent, important, or of very common use;

(b) Two-letter signals for General Signal Code, Chapter 2, Pages 29 through 104;

(c) Three-letter signals beginning with "M" for Medical Signal Code, Chapter 3, pages 107 through 135.

3. The Code follows the basic principle that each signal should have a complete meaning. This principle is followed throughout the Code; in certain cases complements are used, where necessary to supplement the available groups.

4. Complements express:

(a) Variations in the meaning of the basic signal.

Examples:

"CP" = "I am (or vessel indicated is) proceeding to your assistance."

"CP 1" = "SAR aircraft is coming to your assistance."

(b) Questions concerning the same basic subject or basic signal.

Examples:

"DY" = "Vessel (name or identity signal) has sunk in lat . . . long. . . .".

"DY 4" = "What is the depth of water where vessel sank?"

(c) Answers to a question or request made by the basic signal.

Examples:

"HX" = "Have you received any damage in collision?"

"HX 1" = "I have received serious damage above the waterline".

(d) Supplementary, specific or detailed information.

Examples:

"IN" = "I require a diver".

"IN 1" = "I require a diver to clear propeller".

5. Complements appearing in the text more than once have been grouped in three tables. These tables should be used only as and when specified in the text of the signals.

6. Text in brackets indicates:

(a) an alternative, e.g.:". . . (or survival craft). . .";

(b) information which may be transmitted if it is required or if it is available, e.g.: ". . . (position to be indicated if necessary)";

(c) an explanation of the text.

7. The material is classified according to subject and meaning. Extensive cross referencing of the signals in the right-hand column is used to facilitate coding.

CHAPTER 1

SECTION 2: DEFINITIONS

For the purpose of this Code the following terms shall have the meanings defined below:

Visual signaling is any method of communication, the transmission of which is capable of being seen.

Sound signaling is any method of passing Morse signals by means of siren, whistle, foghorn, bell, or other sound apparatus.

Originator is the authority who orders a signal to be sent.

Identity signal or call sign is the group of letters and figures assigned to each station by its administration.

Station means a ship, aircraft, survival craft, or any place at which communications can be effected by any means.

Station of origin is that station where the originator submits a signal for transmission, irrespective of the method of communication employed.

Transmitting station is the station by which a signal is actually being made.

Addressee is the authority to whom a signal is addressed.

Station of destination is that station in which the signal is finally received by the addressee.

Receiving station is the station by which a signal is actually being read.

Procedure denotes the rules drawn up for the conduct of signaling.

Procedure signal is a signal designed to facilitate the conduct of signaling. (See Chapter 1, Section 10, Pages 17, 20, and 21.)

Time of origin is the time at which a signal is ordered to be made.

Group denotes more than one continuous letter and/or numeral which together compose a signal.

A **numeral group** consists of one or more numerals.

A **hoist** consists of one or more groups displayed from a single halyard. A hoist or signal is said to be **at the dip** when it is hoisted about half of the full extent of the halyards. A hoist or signal is said to be **close up** when it is hoisted to the full extent of the halyards.

Tackline is a length of halyard about 2 m (6 ft.) long, used to separate each group of flags.

CHAPTER 1

SECTION 3: METHODS OF SIGNALING

1. The methods of signaling which may be used are:
   (a) Flag signaling, the flags used being those shown on the back cover.
   (b) Flashing light signaling, using the Morse symbols shown in Chapter 1, Section 10, Page 17.
   (c) Sound signaling, using the Morse symbols shown in Chapter 1, Section 10 Page 17.
   (d) Voice over a loud hailer.
   (e) Radiotelegraphy.
   (f) Radiotelephony.
   (g) Morse signaling by hand flags or arms.

Flag signaling

2. A set of signal flags consists of twenty-six alphabetical flags, ten numeral pennants, three substitutes, and the answering pennant. Detailed instructions for signaling by flags are given in Chapter 1, Section 5, Pages 9 and 10.

Flashing light and sound signaling

3. The Morse symbols representing letters, numerals, etc., are expressed by dots and dashes which are signaled either singly or in combination. The dots and dashes and spaces between them should be made to bear the following ratio, one to another, as regards their duration:
   (a) A dot is taken as the unit;
   (b) A dash is equivalent to three units;
   (c) The space of time between any two elements of a symbol is equivalent to one unit; between two complete symbols it is equivalent to three units; and between two words or groups it is equivalent to seven units.
4. In flashing light and sound signaling, while generally obeying the instructions laid down here, it is best to err on the side of making the dots rather shorter in their proportion to the dashes as it then makes the distinction between the elements plainer. The standard rate of signaling by flashing light is to be regarded as forty letters per minute. Detailed instructions for signaling by flashing light and sound are given in Chapter 1, Sections 6 and 7, Pages 11 through 13.

Voice over a loud hailer

5. Whenever possible plain language should be used but where a language difficulty exists groups from the International Code of Signals could be transmitted using the phonetic spelling tables.

Radiotelegraphy and radiotelephony

6. When radiotelegraphy or radiotelephony is used for the transmission of signals, operators should comply with the Radio Regulations of the International Telecommunication Union then in force. (See Radiotelephony in Chapter 1, Section 8, Page 14.)

CHAPTER 1

SECTION 4: GENERAL INSTRUCTIONS

Originator and addressee of message

1. Unless otherwise indicated all signals between vessels are made from the Master of the vessel of origin to the Master of the vessel of destination.

Identification of ships and aircraft

2. Identity signals for ships and aircraft are allocated on an international basis. The identity signal may therefore indicate the nationality of a ship or aircraft.

Use of identity signals

3. Identity signals may be used for two purposes:
   (a) to speak to, or call, a station;
   (b) to speak of, or indicate, a station.
   Examples:
   "YP LABC" = "I wish to communicate with vessel LABC by . . ." (Complements Table 1, Chapter 2, Section 10, Page 104).
   "HY 1 LABC" = "The vessel LABC with which I have been in collision has resumed her voyage".

Names of vessels and/or places

4. Names of vessels and/or places are to be spelled out.
   Example:
   "RV Gibraltar" = "You should proceed to Gibraltar".

How to signal numbers

5. Instructions for signaling numbers:
   (a) Numbers are to be signaled as follows:
      (i) Flag signaling: by the numeral pennants of the Code.
      (ii) Flashing light or sound signaling: usually by the numerals in the Morse Code; they may also be spelled out.
      (iii) Radiotelephony or loud hailer: by the Code words of the Figure Spelling Table in Chapter 1, Section 10, Page 19.
   (b) Figures which form part of the basic signification of a signal are to be sent together with the basic group.
      Examples:
      "DI 20" = "I require boats for 20 persons."
      "FJ 2" = "Position of accident (or survival craft) is marked by sea marker".
   (c) A decimal point between numerals is to be signaled as follows:
      (i) Flag signaling: by inserting the answering pennant where it is desired to express the decimal point.
      (ii) Flashing light and sound signaling: by "decimal point" signal **"AAA"**.
      (iii) Voice: by use of the word "DECIMAL" as indicated in the Figure Spelling Table.
   (d) Wherever the text allows depths, etc., to be signaled in feet or in meters, the figures should be followed by **"F"** to indicate feet or by **"M"** to indicate meters.

Azimuth or bearing

6. They are to be expressed in three figures denoting degrees from 000 to 359, measured clockwise. If there is any possibility of confusion, they should be preceded by the letter **"A"**. They are always to be true unless expressly stated to be otherwise in the context.
   Examples:
   "LW 005" = "I receive your transmission on bearing 005°".
   "LT A120 T1540" = "Your bearing from me is 120° at (local time) 1540".

Course

7. Course is to be expressed in three numerals denoting degrees from 000 to 359, measured clockwise. If there is any possibility of confusion, they should be preceded by the letter "**C**". They are always to be true unless expressly stated to be otherwise in the context.

Examples:
"MD 025" = "My course is 025°".
"GR C240 S18" = "Vessel coming to your rescue is steering course 240°, speed 18 knots".

Date

8. Dates are to be signaled by two, four, or six numerals preceded by the letter "**D**". The first two numerals indicate the day of the month. When they are used alone they refer to the current month.

Example:
"D15" transmitted on the 15th or any other date in April means "15 April".

The two numerals which follow indicate the month of the year.
Example:
"D1504" means "15 April".

Where necessary the year may be indicated by two further numerals.
Example:
"D181063" means "18 October 1963".

Latitude

9. Latitude is expressed by four figures preceded by the Letter "**L**". The first two figures denote the degrees and the last two the minutes. The letters "**N**" (North) or "**S**" (South) follow if they are needed; however, for reasons of simplicity they may be omitted if there is no risk of confusion.

Example:
"L3740S" = "Latitude 37°40'S".

Longitude

10. Longitude is expressed by four or, if necessary, five figures preceded by the letter "**G**". The first two (or three) figures denote the degrees and the last two the minutes. When the longitude is more than 99°, no confusion will normally arise if the figure indicating hundreds of degrees is omitted. However, where it is necessary to avoid confusion the five figures should be used. The letters "**E**" (East) or "**W**" (West) follow if they are needed, otherwise they may be omitted, as in the case of latitude.

Example:
"G13925E" = "Longitude 139°25' E".
A signal requiring the indication of position to complete its signification should be signaled as follows:
"CH L2537N G4015W" = "Vessel indicated is reported as requiring assistance in lat 25°37' N, long 40°15' W".

Distance

11. Figures preceded by the letter "**R**" indicate distance in nautical miles.
Example:
"OV A080 R10" = "Mine(s) is (are) believed to be bearing 080° from me, distance 10 miles".
The letter "**R**" may be omitted if there is no possibility of confusion.

Speed

12. Speed is indicated by figures preceded by:
(a.)   the letter "**S**" to denote speed in knots, or
(b.)   the letter "**V**" to denote speed in kilometers per hour.

Examples:
"BQ S300" = "The speed of my aircraft in relation to the surface of the earth is 300 knots".
"BQ V300" = "The speed of my aircraft in relation to the surface of the earth is 300 kilometers per hour".

Time

13. Times are to be expressed in four figures, of which the first two denote the hour, from 00 (midnight) up to 23 (11 p.m.), and the last two denote the minutes (from 00 to 59). The figures are preceded by:
  (a) the letter **"T"** indicating "Local time", or
  (b) the letter **"Z"** indicating "Greenwich Mean Time".
  Examples:
  "BH T1045 L2015N G3840W C125" = "I sighted an aircraft at local time 1045 in lat 20°15' N, long 38°40' W flying on course 125°".
  "RX Z0830" = "You should proceed at GMT 0830".

Time of origin

14. The time of origin may be added at the end of the text. It should be given to the nearest minute and expressed by four figures. Apart from indicating at what time a signal originated, it also serves as a convenient reference number.

Communication by local signal codes

15. If a vessel or a coast station wishes to make a signal in a local code, the signal **"YV 1"** = "The groups which follow are from the local code" should precede the local signal, if it is necessary, in order to avoid misunderstanding.

CHAPTER 1

SECTION 5: FLAG SIGNALING

1. As a general rule only one hoist should be shown at a time. Each hoist or group of hoists should be kept flying until it has been answered by the receiving station (see paragraph 3). When more groups than one are shown on the same halyard they must be separated by a tackline. The transmitting station should always hoist the signal where it can be most easily seen by the receiving station, that is, in such a position that the flags will blow out clear and be free from smoke.

How to call

2. The identity signal of the station(s) addressed is to be hoisted with the signal (see Chapter 1, Section 4, Paragraph 3, Page 6). If no identity signal is hoisted it will be understood that the signal is addressed to all stations within visual signaling distance. If it is not possible to determine the identity signal of the station to which it is desired to signal, the group **"VF"** = "You should hoist your identity signal" or **"CS"** = "What is the name or identity signal of your vessel (or station)?" should be hoisted first; at the same time the station will hoist its own identity signal. The group **"YQ"** = "I wish to communicate by . . . (Complements Table 1, Chapter 2, Section 10, Page 104) with vessel bearing . . . from me" can also be used.

How to answer signals

3. All stations to which signals are addressed or which are indicated in signals are to hoist the answering pennant at the dip as soon as they see each hoist and close up immediately, when they understand it; it is to be lowered to the dip as soon as the hoist is hauled down at the transmitting station, being hoisted close up again as soon as the next hoist is understood.

How to complete a signal

4. The transmitting station is to hoist the answering pennant singly after the last hoist of the signal to indicate that the signal is completed. The receiving station is to answer this in a similar manner to all other hoists (see paragraph 3 on this page).

How to act when signals are not understood

5. If the receiving station cannot clearly distinguish the signal made to it, it is to keep the answering pennant at the dip. If it can distinguish the signal but cannot understand the meaning of it, it can hoist the following signals: **"ZQ"** = "Your signal appears incorrectly coded. You should check and repeat the whole", or **"ZL"** = "Your signal has been received but not understood".

The use of substitutes

6. The use of substitutes is to enable the same signal flag, either alphabetical flag or numeral pennant, to be repeated one or more times in the same group, in case only one set of flags is carried on board. The first substitute always repeats the uppermost signal flag of that class of flags which immediately precedes the substitute. The second substitute always repeats the second and the third substitute repeats the third signal flag, counting from the top of that class of flags which immediately precedes them. No substitute can ever be used more than once in the same group. The answering pennant when used as a decimal point is to be disregarded in determining which substitute to use.

> *Example:*
> The signal **"VV"** would be made as follows:
> **V**
> **first substitute**
> The number **"1100"** would be made by numeral pennants as follows:
> **1**
> **first substitute**
> **0**
> **third substitute**
> The signal **"L 2330"** would be made as follows:
> **L**
> **2**
> **3**
> **second substitute**
> **0**

In this case, the second substitute follows a numeral pennant and therefore it can only repeat the second numeral in the group.

How to spell

7. Names in the text of a signal are to be spelled out by means of the alphabetical flags. The signal "**YZ**" = "The words which follow are in plain language" can be used, if necessary.

Use of the Code pennant by ships of war

8. When a ship of war wishes to communicate with a merchant vessel she will hoist the Code pennant in a conspicuous position, and keep it flying during the whole of the time the signal is being made.

CHAPTER 1

SECTION 6: FLASHING LIGHT SIGNALING

1. A signal made by flashing light is divided into the following parts:
   (a) The **call**.—It consists of the general call or the identity signal of the station to be called. It is answered by the answering signal.
   (b) The **identity**.—The transmitting station makes **"DE"** followed by its identity signal or name. This will be repeated back by the receiving station which then signals its own identity signal or name. This will also be repeated back by the transmitting station.
   (c) The **text**.—This consists of plain language or Code groups. When Code groups are to be used they should be preceded by the signal **"YU"**. Words of plain language may also be in the text, when the signal includes names, places, etc. Receipt of each word or group is acknowledged by **"T"**.
   (d) The **ending**.—It consists of the ending signal **"$\overline{AR}$"** which is answered by **"R"**.
2. If the entire text is in plain language the same procedure is to be followed. The call and identity may be omitted when two stations have established communications and have already exchanged signals.
3. A list of procedure signals appears in Chapter 1, Section 10, Pages 20 and 21. Although the use of these signals is self-explanatory, the following notes might be found useful:
   (a) The **General call signal** (or call for unknown station) **"$\overline{AA}$ $\overline{AA}$ $\overline{AA}$"** etc., is made to attract attention when wishing to signal to all stations within visual signaling distance or to a station whose name or identity signal is not known. The call is continued until the station addressed answers.
   (b) The **Answering signal "TTTT"** etc., is made to answer the call and it is to be continued until the transmitting station ceases to make the call. The transmission starts with the **"DE"** followed by the name or identity signal of the transmitting station.
   (c) The letter **"T"** is used to indicate the receipt of each word or group.
   (d) The **Erase signal "EEEEEE"** etc., is used to indicate that the last group or word was signaled incorrectly. It is to be answered with the erase signal. When answered, the transmitting station will repeat the last word or group which was correctly signaled and then proceed with the remainder of the transmission.
   (e) The **Repeat signal "RPT"** is to be used as follows:
      (i) by the transmitting station to indicate that it is going to repeat ("I repeat"). If such a repetition does not follow immediately after **"RPT"**, the signal should be interpreted as a request to the receiving station to repeat the signal received ("Repeat what you have received");
      (ii) by the receiving station to request for a repetition of the signal transmitted ("Repeat what you have sent");
      (iii) The **Special Repetition signals "AA", "AB", "WA", "WB"**, and **"BN"** are made by the receiving station as appropriate. In each case they are made immediately after the repeat signal **"RPT"**.
      Examples:
      "RPT AB KL"—"Repeat all before group **KL**".
      "RPT BN 'boats' 'survivors' "—"Repeat all between words 'boats' and 'survivors' ".
      If a signal is not understood, or, when decoded, it is not intelligible, the repeat signal is not used. The receiving station must then make the appropriate signal from the Code, e.g., "Your signal has been received but not understood".
   (f) A correctly received **repetition** is acknowledged by the signal **"OK"**. The same signal may be used as an affirmative answer to a question ("It is correct").
   (g) The **Ending signal "$\overline{AR}$"** is used in all cases to indicate the end of a signal or the end of the transmission. The receiving station answers with the signal **"R"** = "Received" or "I have received your last signal".
   (h) The transmitting station makes the signal **"CS"** when **requesting the name or identity signal** of the receiving station.
   (i) The **Waiting signal** or **Period signal "AS"** is to be used as follows:
      (i) When made independently or after the end of a signal it indicates that the other station must wait for further communications (**waiting signal**);
      (ii) When it is inserted between groups it serves to separate them (**period signal**) to avoid confusion.
   (j) The signal **"C"** should be used to indicate an affirmative statement or an affirmative reply to an interrogative signal; the signal **"RQ"** should be used to indicate a question. For a negative reply to an interrogative signal or for a negative statement, the signal **"N"** should be used in visual or sound signaling and the signal **"NO"** should be used for voice or radio transmission.
   (k) When the signals **"N"** or **"NO"**, and **"RQ"** are used to change an affirmative signal into a negative statement or into a

CHAPTER 1.—SIGNALING INSTRUCTIONS

question, respectively, they should be transmitted after the main signal.

Examples:

"CY N" (or **"NO"** as appropriate) = "(Boat(s) is(are) not coming to you." **"CW RQ"** = "Is boat/raft on board?"

The signals **"C"**, **"N"** or **"NO"**, and **"RQ"** cannot be used in conjunction with single-letter signals.

CHAPTER 1

SECTION 7: SOUND SIGNALING

1. Owing to the nature of the apparatus used (whistle, siren, foghorn, etc.) sound signaling is necessarily slow. Moreover, the misuse of sound signaling is of a nature to create serious confusion at sea. Sound signaling in fog should therefore be reduced to a minimum. Signals other than the single letter signals should be used only in extreme emergency and never in frequented navigational waters.

2. The signals should be made slowly and clearly. They may be repeated, if necessary, but at sufficiently long intervals to ensure that no confusion can arise and that one letter signals cannot be mistaken as two-letter groups.

3. Masters are reminded that the one letter signals of the Code, which are marked by an asterisk(*), when made by sound, may only be made in compliance with the requirements of the International Regulations for Preventing Collisions at Sea. Reference is also made to the single letter signals provided for exclusive use between an icebreaker and assisted vessels.

CHAPTER 1

SECTION 8: RADIOTELEPHONY

1. When using the International Code of Signals in cases of language difficulties, the principles of the Radio Regulations of the International Telecommunication Union then in force have to be observed. Letters and figures are to be spelled in accordance with the phonetic spelling tables in Chapter 1, Section 10, Pages 18 through 20.

2. When coast and ship stations are called, the identity signals (call signs) or names shall be used.

Method of calling

3. The call consists of:
   (a) The call sign or name of the station called, not more than three times at each call;
   (b) The group **"DE" (DELTA ECHO)**;
   (c) The call sign or name of the calling station, not more than three times at each call.

Difficult names of stations should also be spelled. After contact has been established, the call sign or name need not be sent more than once.

Form of reply to calls

4. The reply to calls consists of:
   (a) The call sign or name of the calling station, not more than three times;
   (b) The group "DE" **(DELTA ECHO)**;
   (c) The call sign or name of the station called, not more than three times.

Calling all stations in the vicinity

5. The group **"CQ" (CHARLIE QUEBEC)** shall be used, but not more than three times at each call.

6. In order to indicate that Code groups of the International Code of Signals are to follow, the word **"INTERCO"** is to be inserted. Words of plain language may also be in the text when the signal includes names, places, etc. In this case the group **"YZ" (YANKEE ZULU)** is to be inserted if necessary.

7. If the station called is unable to accept traffic immediately, it should transmit the signal **"AS" (ALFA SIERRA)**, adding the duration of waiting times in minutes whenever possible.

8. The receipt of a transmission is indicated by the signal **"R" (ROMEO)**.

9. If the transmission is to be repeated in total or in part, the signal **"RPT" (ROMEO PAPA TANGO)** shall be used, supplemented as necessary by:
   "AA" (ALFA ALFA) = all after . . .
   "AB" (ALFA BRAVO) = all before . . .
   "BN" (BRAVO NOVEMBER) = all between . . .and . . .
   "WA" (WHISKEY ALFA) = word or group after . . .
   "WB" (WHISKEY BRAVO) = word or group before . . .

10. The end of a transmission is indicated by the signal **"AR" (ALFA ROMEO)**.

CHAPTER 1

SECTION 9: SIGNALING BY HAND FLAGS OR ARMS

MORSE SIGNALING BY HAND FLAGS OR ARMS

1. A station which desires to communicate with another station by Morse signaling by hand flags or arms may indicate the requirement by transmitting to that station the signal "**K1**" by any method. The call signal "$\overline{\text{AA}}\ \overline{\text{AA}}\ \overline{\text{AA}}$" may be made instead.
2. On receipt of the call the station addressed should make the answering signal, or, if unable to communicate by this means, should reply with the signal "**YS1**" by any available method.
3. The call signal "$\overline{\text{AA}}\ \overline{\text{AA}}\ \overline{\text{AA}}$" and the signal "**T**" should be used respectively by the transmitting station and the addressed station.
4. Normally both arms should be used for this method of transmission but in cases where this is difficult or impossible one arm can be used.
5. All signals will end with the ending signal "$\overline{\text{AR}}$".

TABLE OF MORSE SIGNALING BY HAND FLAGS OR ARMS

1 Raising both hand flags or arms	2 Spreading out both hand flags or arms at shoulder level
"Dot"	"Dash"
3 Hand flags or arms brought before the chest	4 Hand flags or arms kept at 45° away from the body downwards
Separation of "dots" and/or "dashes"	Separation of letters, groups or words

5 Circular motion of hand flags or arms over the head

Erase signals, if made by the transmitting station.

Request for repetition if by the receiving station.

Note: The space of time between dots and dashes and between letters, groups, or words should be such as to facilitate correct reception.

CHAPTER 1

SECTION 10: MORSE SYMBOLS—PHONETIC TABLES—PROCEDURE SIGNALS

MORSE SYMBOLS

ALPHABET

A	• —	N	— •
B	— • • •	O	— — —
C	— • — •	P	• — — •
D	— • •	Q	— — • —
E	•	R	• — •
F	• • — •	S	• • •
G	— — •	T	—
H	• • • •	U	• • —
I	• •	V	• • • —
J	• — — —	W	• — —
K	— • —	X	— • • —
L	• — • •	Y	— • — —
M	— —	Z	— — • •

NUMERALS

1	• — — — —	6	— • • • •
2	• • — — —	7	— — • • •
3	• • • — —	8	— — — • •
4	• • • • —	9	— — — — •
5	• • • • •	0	— — — — —

PROCEDURE SIGNALS

AR	• — • — •	AAA	• — • — • —
AS	• — • • •		

PHONETIC TABLES

For the pronunciation of letters and figures by radiotelephony or by voice over a loud hailer.

LETTER SPELLING TABLE

Letter	Code Word	Pronunciation
A	Alfa	**AL** FAH
B	Bravo	**BRAH** VOH
C	Charlie	**CHAR** LEE (or **SHAR** LEE)
D	Delta	**DELL** TAH
E	Echo	**ECK** OH
F	Foxtrot	**FOKS** TROT
G	Golf	GOLF
H	Hotel	HOH **TELL**
I	India	**IN** DEE AH
J	Juliett	**JEW** LEE **ETT**
K	Kilo	**KEY** LOH
L	Lima	**LEE** MAH
M	Mike	MIKE
N	November	NO **VEM** BER
O	Oscar	**OSS** CAH
P	Papa	PAH **PAH**
Q	Quebec	KEH **BECK**
R	Romeo	**ROW** ME OH
S	Sierra	SEE **AIR** RAH
T	Tango	**TANG** GO
U	Uniform	**YOU** NEE FORM (or **OO** NEE FORM)
V	Victor	**VIK** TAH
W	Whiskey	**WISS** KEY
X	X-ray	**ECKS** RAY
Y	Yankee	**YANG** KEY
Z	Zulu	**ZOO** LOO

Note: The **Boldfaced** syllables are emphasized.

FIGURE SPELLING TABLE

Figure or Mark to be Transmitted	Code Word	Pronunciation
0	NADAZERO	NAH-DAH-ZAY-ROH
1	UNAONE	OO-NAH-WUN
2	BISSOTWO	BEES-SOH-TOO
3	TERRATHREE	TAY-RAH-TREE
4	KARTEFOUR	KAR-TAY-FOWER
5	PANTAFIVE	PAN-TAH-FIVE
6	SOXISIX	SOK-SEE-SIX
7	SETTESEVEN	SAY-TAY-SEVEN
8	OKTOEIGHT	OK-TOH-AIT
9	NOVENINE	NO-VAY-NINER
DECIMAL POINT	DECIMAL	DAY-SEE-MAL
FULL STOP	STOP	STOP

Note: Each syllable should be equally emphasized. The second component of each Code word is the Code word used in the Aeronautical Mobile Service.

PROCEDURE SIGNALS

A bar over the letters composing a signal denotes that the letters are to be made as one symbol.

1. Signals for voice transmissions (radiotelephony or loud hailer):

Signal	Pronunciation	Meaning
Interco	IN-TER-CO	International Code group(s) follows(s).
Stop	STOP	Full stop.
Decimal	DAY-SEE-MAL	Decimal point.
Correction	KOR-REK-SHUN	Cancel my last word or group. The correct word or group follows.

2. Signals for flashing light transmission:

$\overline{AA}\ \overline{AA}\ \overline{AA}$ etc.	Call for unknown station or general call.
$\overline{EEEEEE}$ etc.	Erase signal.
$\overline{AAA}$	Full stop or decimal point.
$\overline{TTTT}$ etc.	Answering signal.
T	Word or group received.

3. Signals for flags, radiotelephony, and radiotelegraphy transmissions:

CQ	Call for unknown station(s) or general call to all stations. *Note:* When this signal is used in voice transmission, it should be pronounced in accordance with the letter spelling table.

4. Signals for use where appropriate in all forms of transmission:

AA	"All after . . ." (used after the "Repeat signal" (**RPT**)) means "Repeat all after . . .".
AB	"All before . . ." (used after the "Repeat signal" (**RPT**)) means "Repeat all before . . .".
$\overline{AR}$	Ending signal or End of Transmission or signal.
$\overline{AS}$	Waiting signal or period.
BN	"All between . . . and . . ."(used after the "Repeat signal" (**RPT**)) means "Repeat all between . . . and . . .".
C	Affirmative—**YES** or "The significance of the previous group should be read in the affirmative".
CS	"What is the name or identity signal of your vessel (or station)?".
DE	"From . . ." (used to precede the name or identity signal of the calling station).
K	"I wish to communicate with you" or "Invitation to transmit".

NO	Negative—**NO** or "The significance of the previous group should be read in the negative". When used in voice transmission the pronunciation should be **"NO"**.
OK	Acknowledging a correct repetition or "It is correct".
RQ	Interrogative, or, "The significance of the previous group should be read as a question".
R	"Received" or "I have received your last signal".
RPT	Repeat signal "I repeat" or "Repeat what you have sent" or "Repeat what you have received".
WA	"Word or group after . . ." (used after the "Repeat signal" (**RPT**)) means "Repeat word or group after . . .".
WB	"Word or group before . . ." (used after the "Repeat signal" (**RPT**)) means "Repeat word or group before . . .".

Notes:
  (a) The procedure signals **"C", "N", "NO"**, and **"RQ"** cannot be used in conjunction with single-letter signals.
  (b) Signals on COMMUNICATIONS appear in Chapter 2, Section 8, Pages 100 through 102.
  (c) When these signals are used by voice transmission the letters should be pronounced in accordance with the letter-spelling table, with the exception of **"NO"** which in voice transmission should be pronounced as **"NO"**.

SINGLE LETTER SIGNALS

May be made by any method of signaling.

See Note 1 for those marked by an asterisk (*)

A   I have a diver down; keep well clear at slow speed.

*B   I am taking in, or discharging, or carrying dangerous goods.

*C   Yes (affirmative or "The significance of the previous group should be read in the affirmative").

*D   Keep clear of me; I am maneuvering with difficulty.

*E   I am altering my course to starboard.

F   I am disabled; communicate with me.

*G   I require a pilot. When made by fishing vessels operating in close proximity on the fishing grounds it means: "I am hauling nets".

*H   I have a pilot on board.

*I   I am altering my course to port.

J   I am on fire and have dangerous cargo on board: keep well clear of me, or I am leaking dangerous cargo.

K   I   wish to communicate with you.

L   You should stop your vessel instantly.

M   My vessel is stopped and making no way through the water.

N   No (negative or "The significance of the previous group should be read in the negative"). This signal may be given only visually or by sound. For voice or radio transmission the signal should be **"NO"**.

O   Man overboard.

P   **In harbor.**—All persons should report on board as the vessel is about to proceed to sea.
At sea.—It may be used by fishing vessels to mean: "My nets have come fast upon an obstruction". It may also be used as a sound to mean: "I require a pilot".

Q   My vessel is "healthy" and I request free pratique.

*S   I am operating astern propulsion.

*T   Keep clear of me; I am engaged in pair trawling.

U   You are running into danger.

V   I require assistance.

W   I require medical assistance.

X   Stop carrying out your intentions and watch for my signals.

Y   I am dragging my anchor.

*Z   I require a tug. When made by fishing vessels operating in close proximity on the fishing grounds it means: "I am shooting nets".

Notes: 1. Signals of letters marked by an asterisk (*) when made by sound may only be made in compliance with the requirements of the International Regulations for Preventing Collisions at Sea, 1972.
2. Signals **"K"** and **"S"** have special meanings as landing signals for small boats with crews or persons in distress. (International Convention for the Safety of Life at Sea, 1974, Chapter V, Regulation 16.)

SINGLE LETTER SIGNALS WITH COMPLEMENTS
May be made by any method of signaling.

A—with three numerals... AZIMUTH or BEARING.

C—with three numerals... COURSE.

D—with two, four, or six numerals .... DATE.

G—with four or five numerals... LONGITUDE (the last two numerals denote minutes and the rest degrees).

K—with one numeral .... I wish to COMMUNICATE with you by . . . (Complements Table 1, Chapter 2, Section 10, Page 104).

L—with four numerals .... LATITUDE (the first two denote degrees and the rest minutes).

R—with one or more numerals .... DISTANCE in nautical miles.

S—with one or more numerals .... SPEED in knots.

T—with four numerals .... LOCAL TIME (the first two denote hours and the rest minutes).

V—with one or more numerals .... SPEED in kilometers per hour.

Z—with four numerals .... GMT (the first two denote hours and the rest minutes).

Z—with one numeral... To call or address shore visual stations (Numeral to be approved by local port authority).

AZIMUTH or BEARING... **A** with three numerals.

COMMUNICATE, I wish to communicate with you by . . . (Complements Table 1, Chapter 2, Section 10, Page 104) ... **K** with one numeral.

COURSE.. **C** with three numerals.

DATE .... **D** with two, four, or six numerals.

DISTANCE in nautical miles .... **R** with one or more numerals.

GMT (the first two denote hours and the rest minutes) .... **Z** with four numerals.

LATITUDE (the first two denote degrees and the rest minutes) .... **L** with four numerals.

LONGITUDE (the last two numerals denote minutes and the rest degrees) .... **G** with four or five numerals.

LOCAL TIME (the first two denote hours and the rest minutes) .... **T** with four numerals.

SPEED in kilometers per hour.. **V** with one or more numerals.

SPEED in knots .... **S** with one or more numerals.

SINGLE LETTER SIGNALS BETWEEN ICEBREAKER AND ASSISTED VESSELS

The following single letter signals, when made between an icebreaker and assisted vessels, have only the significations given in this table and are only to be made by sound, visual, or radiotelephony signals.

WM   Icebreaker support is now commencing. Use special icebreaker support signals and keep continuous watch for sound, visual, or radiotelephony signals.

WO   Icebreaker support is finished. Proceed to your destination.

Code Letters or Numerals	Icebreaker	Assisted Vessel(s)
A • –	Go ahead (proceed along the ice channel).	I am going ahead (I am proceeding along the ice channel).
G – – •	I am going ahead; follow me.	I am going ahead; I am following you.
J • – – –	Do not follow me (proceed along the ice channel).	I will not follow you (I will proceed along the ice channel).
P • – – •	Slow down.	I am slowing down.
N – •	Stop your engines.	I am stopping my engines.
H • • • •	Reverse your engines.	I am reversing my engines.
L • – • •	You should stop your vessel instantly.	I am stopping my vessel.
4 • • • • –	Stop. I am icebound.	Stop. I am icebound.
Q – – • –	Shorten the distance between vessels.	I am shortening the distance.
B – • • •	Increase the distance between vessels.	I am increasing the distance.
5 • • • • •	Attention.	Attention.
Y – • – –	Be ready to take (or cast off) the towline.	I am ready to take (or cast off) the towline.

Notes:

1. The signal **"K"**– • – by sound or light may be used by an icebreaker to remind ships of their obligation to listen continuously on their radio.

2. If more than one vessel is assisted, the distances between vessels should be as constant as possible. Watch speed of your own vessel and vessel ahead; should speed of your own vessel go down, give attention signal to the following vessel.

3. The use of these does not relieve any vessel from complying with the International Regulations for Preventing Collisions at Sea.

4.	• • – • •	Stop your headway (given only to a ship in an ice-channel ahead of and approaching or going away from icebreaker).	I am stopping headway.

Note: This signal should not be made by radiotelephone.

Single-letter signals which may be used during icebreaking operations:

* **E** •   I am altering my course to starboard.

* **I** • •   I am altering my course to port.

* **S** • • •   I am operating astern propulsion.

M – –   My vessel is stopped and making no way through the water.

Notes:

1. Signals of letters marked by an asterisk*, when made by sound, may only be made in compliance with the requirements of the International Regulations for Preventing Collisions at Sea.

2. Additional signals for icebreaking support can be found in Chapter 2, Section 6, Page 93 and 94.

CHAPTER **2**

CHAPTER 2
GENERAL SIGNAL CODE

CHAPTER 2

SECTION 1: DISTRESS—EMERGENCY

Code   *Meaning*   *Cross Reference*

ABANDON

*AA   Repeat all after…

*AB   Repeat all before…

AC   I am abandoning my vessel.

AD   I am abandoning my vessel which has suffered a nuclear accident and is a possible source of radiation danger.

AE   I must abandon my vessel.

    AE 1   I (or crew of vessel indicated) wish to abandon my (or their) vessel, but have not the means.

    AE 2   I shall abandon my vessel unless you will remain by me, ready to assist.

AF   I do not intend to abandon my vessel.

    AF 1   Do you intend to abandon your vessel?

AG   You should abandon your vessel as quickly as possible.

AH   You should not abandon your vessel.

AI   Vessel (indicated by position and/or name or identity signal if necessary) will have to be abandoned.

  *   Procedural signals for repetition.

ACCIDENT—DOCTOR—INJURED/SICK

Accident

AJ   I have had a serious nuclear accident and you should approach with caution.

AK   I have had a nuclear accident on board.

I am abandoning my vessel which has suffered a nuclear accident and is a possible source of radiation danger . . . . **AD**

I am proceeding to the position of accident . . . . **SB**

I am proceeding to the position of accident at full speed. Expect to arrive at time indicated . . **FE**

Are you proceeding to the position of accident? If so, when do you expect to arrive? . . . . **FE 1**

You should steer course… (or follow me) to reach position of accident . . . . **FL**

I am circling over the area of accident . . . . **BJ**

An aircraft is circling over the area of accident . . . . **BJ 1**

Position of accident (or survival craft) is marked . . . . **FJ**

Code		Meaning	Cross Reference

Accident

		Position of accident (or survival craft) is marked by flame or smoke float	**FJ 1**
		Position of accident (or survival craft) is marked by sea marker	**FJ 2**
		Position of accident (or survival craft) is marked by sea marker dye	**FJ 3**
		Position of accident (or survival craft) is marked by radiobeacon	**FJ 4**
		Position of accident (or survival craft) is marked by wreckage	**FJ 5**
		Is position of accident (or survival craft) marked?	**FK**
		I have searched area of accident but have found no trace of derelict or survivors	**GC 2**
		Man overboard. Please take action to pick him up (position to be indicated if necessary)	**GW**

Doctor

AL		I have a doctor on board.	
AM		Have you a doctor?	
AN		I need a doctor.	
	AN 1	I need a doctor; I have severe burns.	
	AN 2	I need a doctor; I have radiation casualties.	
		I require a helicopter urgently, with a doctor	**BR 2**
		Helicopter is coming to you now (or at time indicated) with a doctor	**BT 2**

Injured/Sick

AO		Number of injured and/or dead not yet known.	
	AO 1	How many injured?	
	AO 2	How many dead?	
AP		I have… (number) casualties.	
AQ		I have injured/sick person (or number of persons indicated) to be taken off urgently. I cannot alight but I can lift injured/sick person.	**AZ 1**
		You cannot alight on the deck; can you lift injured/sick person?	**BA 2**
		I require a helicopter urgently to pick up injured/sick person	**BR 3**
		You should send a helicopter/boat with a stretcher	**BS**
		A helicopter/boat is coming to take injured/sick	**BU**
AT		You should send injured/sick persons to me.	

Code		*Meaning*	*Cross Reference*

AIRCRAFT-HELICOPTER

Alight-Landing

AU		I am forced to alight near you (or in position indicated).	
AV		I am alighting (in position indicated if necessary) to pick up crew of vessel/aircraft	
AW		Aircraft should endeavor to alight where flag is waved or light is shown.	
AX		You should train your searchlight nearly vertical on a cloud, intermittently if possible, and, if my aircraft is seen, deflect the beam upwind and on the water to facilitate my landing.	
	AX 1	Shall I train my searchlight nearly vertical on a cloud, intermittently if possible, and, if your aircraft is seen, deflect the beam upwind and on the water to facilitate your landing?	
AY		I will alight on your deck; (you should steer course… speed… knots).	
AZ		I cannot alight but I can lift crew.	
	AZ 1	I cannot alight but I can lift injured/sick person.	
BA		You cannot alight on the deck.	
	BA 1	You cannot alight on the deck; can you lift crew?	
	BA 2	You cannot alight on the deck; can you lift injured/sick person?	
BB		You may alight on my deck.	
	BB 1	You may alight on my deck; I am ready to receive you forward.	
	BB 2	You may alight on my deck; I am ready to receive you amidship.	
	BB 3	You may alight on my deck; I am ready to receive you aft.	
	BB 4	You may alight on my deck but I am not yet ready to receive you.	

Communications

BC		I have established communications with the aircraft in distress on 2182 kHz.	
	BC 1	Can you communicate with the aircraft?	
BD		I have established communications with the aircraft in distress on… kHz.	
BE		I have established communications with the aircraft in distress on… MHz.	

Ditched-Disabled-Afloat

BF		Aircraft is ditched in position indicated and requires immediate assistance.	
		I sighted disabled aircraft in lat… long… at time indicated	**DS**
BG		Aircraft is still afloat.	

CHAPTER 2.—GENERAL SIGNAL CODE

Code		Meaning	Cross Reference

Flying

BH I sighted an aircraft at time indicated in lat… long… flying on course…

    BH 1 Aircraft was flying at high altitude.

    BH 2 Aircraft was flying at low altitude.

BI I am flying to likely position of vessel in distress.

    BI 1 I am flying at low altitude near the vessel.

BJ I am circling over the area of accident.

    BJ 1 An aircraft is circling over the area of accident.

BK You are overhead.

    BK 1 Am I overhead?

BL I am having engine trouble but am continuing flight.

Parachute

BM You should parachute object to windward. Mark it by smoke or light signal.

    BM 1 I am going to parachute object to windward, marking it by smoke or light signal.

    BM 2 I am going to parachute equipment.

    BM 3 Inflatable raft will be dropped to windward by parachute.

*BN Repeat all between… and…

BO We are going to jump by parachute.

* Procedural signal for repetition.

Search—Assistance

BP Aircraft is coming to participate in search. Expected to arrive over the area of accident at time indicated.

The search area of the aircraft is between lat… and… , and long… and. . . . . **FU**

Search by aircraft/helicopter will be discontinued because of unfavorable conditions . . . . **FV**

SAR aircraft is coming to your assistance . . . . **CP 1**

Speed

BQ The speed of my aircraft in relation to the surface of the earth is… (knots or kilometers per hour).

Code	*Meaning*	*Cross Reference*

Speed

BQ 1   What is the speed of your aircraft in relation to the surface of the earth?

Helicopter

BR   I require a helicopter urgently.

      BR 1   I require a helicopter urgently to pick up persons.

      BR 2   I require a helicopter urgently with a doctor.

      BR 3   I require a helicopter urgently to pick up injured/sick person.

      BR 4   I require a helicopter urgently with inflatable raft.

BS   You should send a helicopter/boat with stretcher.

BT   Helicopter is coming to you now (or at time indicated).

      BT 1   Helicopter is coming to you now (or at time indicated) to pick up persons.

      BT 2   Helicopter is coming to you now (or at time indicated) with a doctor.

      BT 3   Helicopter is coming to you now (or at time indicated) to pick up injured/sick person.

      BT 4   Helicopter is coming to you now (or at time indicated) with inflatable raft.

BU   A helicopter/boat is coming to take injured/sick.

BV   I cannot send a helicopter.

BW   The magnetic course for you to steer towards me (or vessel or position indicated) is… (at time indicated).

BX   The magnetic course for the helicopter to regain its base is…

BY   Will you indicate the magnetic course for me to steer towards you (or vessel or position indicated)?

BZ   Your magnetic bearing from me (or from vessel or position indicated) is… (at time indicated).

CA   What is my magnetic bearing from you (or from vessel or position indicated)?

ASSISTANCE

Required

I am in distress and require immediate assistance . . . . **NC**

CB   I require immediate assistance.

      CB 1   I require immediate assistance; I have a dangerous list.

      CB 2   I require immediate assistance; I have damaged steering gear.

Code		Meaning	Cross Reference

Code *Meaning* *Cross Reference*

Required

	CB 3	I require immediate assistance; I have a serious disturbance on board.	
	CB 4	I require immediate assistance; I am aground.	
	CB 5	I require immediate assistance; I am drifting.	
	CB 6	I require immediate assistance; I am on fire.	
	CB 7	I require immediate assistance; I have sprung a leak.	
	CB 8	I require immediate assistance; propeller shaft is broken.	
CC		I am (or vessel indicated is) in distress in lat… long… (or bearing… from place indicated, distance…) and require immediate assistance (Complements Table 2, Chapter 2, Section 10, Page 104 if required).	
		I require assistance	**V**
CD		I require assistance in the nature of… (Complements Table 2, Chapter 2, Section 10, Page 104).	
		I require medical assistance	**W**
		I request assistance from fishery protection (or fishery assistance) vessel.	**TY**
CE		I will attempt to obtain for you the assistance required.	
		Aircraft is ditched in position indicated and requires immediate assistance	**BF**
CF		Signals from vessel/aircraft requesting assistance are coming from bearing…from me (lat… long… if necessary).	
CG		Stand by to assist me (or vessel indicated).	
	CG 1	I will stand by to assist you (or vessel indicated).	
		Survivors are in bad condition. Medical assistance is urgently required	**HM**
CH		Vessel indicated is reported as requiring assistance in lat… long… (or bearing… from place indicated, distance…).	
	CH 1	Lightvessel (or lighthouse) indicated requires assistance.	
	CH 2	Space ship is down in lat… long… and requires immediate assistance.	
CI		Vessel aground in lat… long… requires assistance.	
CJ		Do you require assistance?	
	CJ 1	Do you require immediate assistance?	
	CJ 2	Do you require any further assistance?	
	CJ 3	What assistance do you require?	
	CJ 4	Can you proceed without assistance?	

Code		Meaning	Cross Reference

Not Required—Declined

| CK | | Assistance is not (or is no longer) required by me (or vessel indicated). | |
| CL | | I offered assistance but it was declined. | |

Given—Not Given

CM		One or more vessels are assisting the vessel in distress.	
	CM 1	Vessel/aircraft reported in distress is receiving assistance.	
CN		You should give all possible assistance.	
	CN 1	You should give immediate assistance to pick up survivors.	
	CN 2	You should send survival craft to assist vessel indicated.	
CO		Assistance cannot be given to you (or vessel/aircraft indicated).	
	CO 1	I cannot give the assistance required.	

Proceeding to Assistance

CP		I am (or vessel indicated is) proceeding to your assistance.	
	CP 1	SAR aircraft is coming to your assistance.	
*CQ		Call for unknown station(s) or general call to all stations.	
CR		I am proceeding to the assistance of vessel (lat… long…).	
*CS		What is the name or identity signal of your vessel (or station)?	
CT		I (or vessel indicated) expect to reach you at time indicated.	
CU		Assistance will come at time indicated.	
	CU 1	I can assist you.	
CV		I am unable to give assistance.	
	CV 1	Will you go to the assistance of vessel indicated (in lat… long…)?	
	CV 2	May I assist you?	
	CV 3	Can you assist me (or vessel indicated)?	
	CV 4	Can you assist?	
		Can you offer assistance? (Complements Table 2, Chapter 2, Section 10, Page 104).	TZ
		I shall abandon my vessel unless you will remain by me, ready to assist	AE 2
		I cannot get the fire under control without assistance	IX 1

Code	*Meaning*	*Cross Reference*

Proceeding to Assistance

I can get the fire under control without assistance	**IY**
Can you get the fire under control without assistance?	**IY 1**
I have placed the collision mat. I can proceed without assistance	**KA 1**
I cannot take you (or vessel indicated) in tow, but I will report you and ask for immediate assistance	**KN 1**
I cannot steer without assistance	**PK**

* Procedural signals.

BOATS—RAFTS

CW		Boat/raft is on board.
	CW 1	Boat/raft is safe.
	CW 2	Boat/raft is in sight.
	CW 3	Boat/raft is adrift.
	CW 4	Boat/raft is aground.
	CW 5	Boat/raft is alongside.
	CW 6	Boat/raft is damaged.
	CW 7	Boat/raft has sunk.
	CW 8	Boat/raft has capsized.
CX		Boats cannot be used.
	CX 1	Boats cannot be used because of prevailing weather conditions.
	CX 2	Boats cannot be used on the starboard side because of list.
	CX 3	Boats cannot be used on the port side because of list.
	CX 4	Boats cannot be used to disembark people.
	CX 5	Boats cannot be used to get alongside.
	CX 6	Boats cannot be used to reach you.
	CX 7	I cannot send a boat.
CY		Boat(s) is(are) coming to you.
	CY 1	Boat/raft is making for the shore.
	CY 2	Boat/raft has reached the shore.

Code		Meaning	Cross Reference
CZ		You should make a lee for the boat(s)/raft(s).	
	CZ 1	You should discharge oil to smooth sea.	
DA		Boat(s)/raft(s) should approach vessel as near as possible to take off persons.	
		A boat/helicopter is coming to take injured/sick	**BU**
DB		Veer a boat or raft on a line.	
DC		Boat should endeavor to land where flag is waved or light is shown.	
DD		Boats are not allowed to come alongside.	
	DD 1	Boats are not allowed to land (after time indicated).	
***DE**		From…	

* Procedural signal used to precede the name or identity signal of the calling station.

Available

Code		Meaning	Cross Reference
DF		I have… (number) serviceable boats.	
DG		I have a motor boat [or… (number) motor boats].	
DH		I have no boat/raft.	
	DH 1	I have no motor boat.	
	DH 2	Have you any boats with radiotelegraph installation or portable radio equipment?	
	DH 3	How many serviceable motor boats have you?	
	DH 4	How many serviceable boats have you?	

Required

Code		Meaning	Cross Reference
DI		I require boats for… (number) persons.	
DJ		Do you require a boat?	

Send

Code		Meaning	Cross Reference
DK		You should send all available boats/rafts.	
	DK 1	You should send back my boat.	
	DK 2	Can you send a boat?	
		You should send a boat/helicopter with stretcher.	**BS**
		You should send survival craft to assist vessel indicated	**CN 2**
		You should stop, or heave to; I am going to send a boat	**SQ 2**

Code		Meaning	Cross Reference

Send

DL		I can send a boat.	
	DL 1	I am sending a boat.	
		I cannot send a boat	**CX 7**

Search

DM		You should search for the boat(s)/raft(s).	
DN		I have found the boat/raft.	
	DN 1	Have you seen or heard anything of the boat/raft?	
DO		Look out for boat/raft in bearing... distance... from me (or from position indicated).	
DP		There is a boat/raft in bearing... distance... from me (or from position indicated).	
DQ		An empty boat/raft has been sighted in lat... long... (or bearing... from place indicated, distance...).	

DISABLED—DRIFTING—SINKING

Disabled

DR		Have you sighted disabled vessel/aircraft in approximate lat... long...?	
DS		I sighted disabled aircraft in lat... long... at time indicated.	
DT		I sighted disabled vessel in lat... long... at time indicated.	
	DT 1	I sighted disabled vessel in lat... long... at time indicated, apparently without a radio.	
		I am disabled; communicate with me.	**F**

Drifting

DU		I am drifting at... (number) knots, towards... degrees.	
DV		I am drifting.	
	DV 1	I am adrift.	
DW		Vessel (name or identity signal) is drifting near lat... long....	
		I require immediate assistance; I am drifting.	**CB 5**
		I am (or vessel indicated is) breaking adrift.	**RC**
		I have broken adrift.	**RC 1**

Code	Meaning	Cross Reference

Sinking

DX	I am sinking (lat… long… if necessary).	
DY	Vessel (name or identity signal) has sunk in lat… long…	
DY 1	Did you see vessel sink?	
DY 2	Where did vessel sink?	
DY 3	Is it confirmed that vessel (name or identity signal) has sunk?	
DY 4	What is the depth of water where vessel sunk?	

DISTRESS

Vessel/Aircraft in Distress

	I am in distress and require immediate assistance	**NC**
DZ	Vessel (or aircraft) indicated appears to be in distress.	
DZ 1	Is vessel (or aircraft) indicated in distress?	
DZ 2	What is the name (or identity signal) of vessel in distress?	
EA	Have you sighted or heard of a vessel in distress? (Approximate position lat… long… or bearing… from place indicated, distance…).	
EA 1	Have you any news of vessel/aircraft reported missing or in distress in this area?	
	I am (or vessel indicated is) in distress in lat… long… (or bearing… from place indicated, distance…) and require immediate assistance (Complements Table 2, Chapter 2, Section 10, Page 104, if required)	**CC**
EB	There is a vessel (or aircraft) in distress in lat… long… (or bearing… distance… from me, or Complements Table 3, Chapter 2, Section 10, Page 104).	
EC	A vessel which has suffered a nuclear accident is in distress in lat… long…	

Distress Signal

*ED	Your distress signals are understood.	
ED 1	Your distress signals are understood; the nearest life-saving station is being informed.	
EF	SOS/MAYDAY has been cancelled.	
EF 1	Has the SOS/MAYDAY been cancelled?	
	I have intercepted SOS/MAYDAY from vessel (name or identity signal) (or aircraft) in position lat… long… at time indicated.	**FF**

Code		Meaning	Cross Reference

Distress Signal

EG   Did you hear SOS/MAYDAY given at time indicated?

    EG 1   Will you listen on 2182 kHz for signals of emergency position-indicating radiobeacons?

    EG 2   I am listening on 2182 kHz for signals of emergency position-indicating radiobeacons.

    EG 3   Have you received the signal of an emergency position-indicating radiobeacon on 2182 kHz?

    EG 4   I have received the signal of an emergency position-indicating radiobeacon on 2182 kHz.

    EG 5   Will you listen on… MHz for signals of emergency position-indicating radiobeacons?

    EG 6   I am listening on… MHz for signals of emergency position-indicating radiobeacons.

    EG 7   Have you received the signal of an emergency position-indicating radiobeacon on… MHz?

    EG 8   I have received the signal of an emergency position-indicating radiobeacon on… MHz.

EJ   I have received distress signal transmitted by coast station indicated.

    EJ 1   Have you received distress signal transmitted by coast station indicated?

EK   I have sighted distress signal in lat… long…

    EK 1   An explosion was seen or heard (position or direction and time to be indicated).

    EK 2   Have you heard or seen distress signal from survival craft?

*   Reference is made to signals prescribed by the International Convention for the Safety of Life at Sea, 1974 (Regulation 16(a), Chapter V) as replies from lifesaving stations or maritime rescue units to distress signals made by a ship or person.

Position of Distress

EL   Repeat the distress position.

    EL 1   What is the position of vessel in distress?

Position given with SOS/MAYDAY from vessel (or aircraft) was lat… long… (or bearing… from place indicated, distance…) . . . .   **FG**

What was the position given with SOS/MAYDAY from vessel (or aircraft)? . . . .   **FG 1**

Position given with SOS/MAYDAY is wrong. The correct position is lat… long… . . . .   **FH**

Position given with SOS/MAYDAY by vessel is wrong. I have her bearing by radio direction finder and can exchange bearings with any other vessel . . . .   **FI**

Survival craft are believed to be in the vicinity of lat… long… . . . .   **GI**

EM   Are there other vessels/aircraft in the vicinity of vessel/aircraft in distress?

Code	Meaning	Cross Reference

Contact or Locate

EN	You should try to contact vessel/aircraft in distress.	
EO	I am unable to locate vessel/aircraft in distress because of poor visibility.	
EP	I have lost sight of you.	
	I have located (or found) wreckage from the vessel/aircraft in distress (position to be indicated if necessary by lat… long… or by bearing… from specified place, and distance…)	GL
EQ	I expect to be at the position of vessel/aircraft in distress at time indicated.	
EQ 1	Indicate estimated time of your arrival at position of vessel/aircraft in distress.	
	I am flying to likely position of vessel in distress	BI
	One or more vessels are assisting the vessel in distress	CM
	Vessel/aircraft reported in distress is receiving assistance	CM 1
	I am proceeding to the assistance of vessel/aircraft in distress in lat… long…	CR
	I have found vessel/aircraft in distress in lat… long…	GF

POSITION

ER	You should indicate your position at time indicated.	
ET	My position at time indicated was lat… long…	
EU	My present position is lat… long… (or bearing… from place indicated, distance…).	
EU 1	What is your present position?	
EV	My present position, course, and speed are lat… long… ,… , knots…	
EV 1	What are your present position, course, and speed?	
EW	My position is ascertained by dead reckoning.	
EW 1	My position is ascertained by visual bearings.	
EW 2	My position is ascertained by astronomical observations.	
EW 3	My position is ascertained by radiobeacons.	
EW 4	My position is ascertained by radar.	
EW 5	My position is ascertained by electronic position-fixing system.	
EX	My position is doubtful.	
EY	I am confident as to my position.	
EY 1	Are you confident as to your position?	

CHAPTER 2.—GENERAL SIGNAL CODE

Code	Meaning	Cross Reference

EZ — Your position according to bearings taken by radio direction finder stations which I control is lat… long… (at time indicated).

EZ 1 — Will you give me my position according to bearings taken by radio direction finder stations which you control?

FA — Will you give me my position?

FB — Will vessels in my immediate vicinity (or in the vicinity of lat… long…) please indicate position, course, and speed.

Position of Distress

FC — You should indicate your position by visual or sound signals.

FC 1 — You should indicate your position by rockets or flares.

FC 2 — You should indicate your position by visual signals.

FC 3 — You should indicate your position by sound signals.

FC 4 — You should indicate your position by searchlight.

FC 5 — You should indicate your position by smoke signal.

FD — My position is indicated by visual or sound signals.

FD 1 — My position is indicated by rockets or flares.

FD 2 — My position is indicated by visual signals.

FD 3 — My position is indicated by sound signals.

FD 4 — My position is indicated by searchlight.

FD 5 — My position is indicated by smoke signal.

I expect to be at the position of vessel/aircraft in distress at time indicated …………… **EQ**

Indicate estimated time of your arrival at position of vessel/aircraft in distress ………… **EQ 1**

Position given with SOS/MAYDAY from vessel (or aircraft) was lat… long… (or bearing… from place indicated, distance…)………………………………………… **FG**

What was position given with SOS/MAYDAY from vessel (or aircraft)?……………… **FG 1**

Position given with SOS/MAYDAY is wrong. The correct position is lat… long… …… **FH**

Position given with SOS/MAYDAY by vessel is wrong. I have her bearing by radio direction finder and can exchange bearings with any other vessel ………………………… **FI**

Position of accident (or survival craft) is marked.………………………… **FJ**

Position of accident (or survival craft) is marked by flame (or smoke float)…………… **FJ 1**

Position of accident (or survival craft) is marked by sea marker…………………… **FJ 2**

Position of accident (or survival craft) is marked by sea marker dye ……………… **FJ 3**

Code	Meaning	Cross Reference

Position of Distress

	Position of accident (or survival craft) is marked by radiobeacon.....................	**FJ 4**
	Position of accident (or survival craft) is marked by wreckage........................	**FJ 5**
	Is position of accident (or survival craft) marked?	**FK**
	You should transmit your identification and series of long dashes or your carrier frequency to home vessel (or aircraft) to your position...	**FQ**
	Shall I home vessel (or aircraft) to my position?..................................	**FQ 1**
	You should indicate position of survivors by throwing pyrotechnic signals	**HT**

SEARCH AND RESCUE

Proceeding to Assistance

	I am proceeding to the assistance of vessel/aircraft in distress (lat... long...)	**CR**
FE	I am proceeding to the position of accident at full speed. Expect to arrive at time indicated.	
FE 1	Are you proceeding to the position of accident? If so, when do you expect to arrive?	
	I am unable to give assistance	**CV**
	Can you assist?...	**CV 4**

Position of Distress or Accident

FF	I have intercepted SOS/MAYDAY from vessel (name or identity signal) (or aircraft) in position lat... long... at time indicated.	
FF 1	I have intercepted SOS/MAYDAY from vessel (name or identity signal) (or aircraft) in position lat... long... at time indicated; I have heard nothing since.	
FG	Position given with SOS/MAYDAY from vessel (or aircraft) was lat... long... (or bearing... from place indicated, distance...).	
FG 1	What was position given with SOS/MAYDAY from vessel (or aircraft)?	
FH	Position given with SOS/MAYDAY is wrong. The correct position is lat... long...	
FI	Position given with SOS/MAYDAY by vessel is wrong. I have her bearing by radio direction finder and can exchange bearings with any other vessel.	
FJ	Position of accident (or survival craft) is marked.	
FJ 1	Position of accident (or survival craft) is marked by flame or smoke float.	
FJ 2	Position of accident (or survival craft) is marked by sea marker.	
FJ 3	Position of accident (or survival craft) is marked by sea marker dye.	

Code		Meaning	Cross Reference

Position of Distress or Accident

	FJ 4	Position of accident (or survival craft) is marked by radiobeacon.	
	FJ 5	Position of accident (or survival craft) is marked by wreckage.	
FK		Is position of accident (or survival craft) marked?	

Information—Instructions

FL		You should steer course... (or follow me) to reach position of accident.	
		Course to reach me is	**MF**
		What is the course to reach you?	**MF1**
FM		Visual contact with vessel is not continuous.	
FN		I have lost all contact with vessel.	
		I have lost sight of you.	**EP**
FO		I will keep close to you.	
	FO 1	I will keep close to you during the night.	
FP		Estimated set and drift of survival craft is... degrees and... knots.	
	FP 1	What is the estimated set and drift of survival craft?	
FQ		You should transmit your identification and series of long dashes or your carrier frequency to home vessel (or aircraft) to your position.	
	FQ 1	Shall I home vessel (or aircraft) to my position?	

Search

FR		I am (or vessel indicated is) in charge of coordinating search.	
	***FR 1**	Carry out search pattern... starting at... hours. Initial course... search speed... knots.	
	***FR 2**	Carry out radar search, ships proceeding in loose line abreast at intervals between ships...miles. Initial course... search speed... knots.	
	***FR 3**	Vessel indicated (call sign or identity signal) is allocated track number...	
	***FR 4**	Vessel(s) indicated adjust interval between ships to... miles.	
	***FR 5**	Adjust track spacing to... miles.	
	***FR 6**	Search speed will now be... knots.	
	***FR 7**	Alter course as necessary to next leg of track now (or at time indicated).	
FS		Please take charge of search in sector stretching between bearings... and... from vessel in distress.	

Code		Meaning	Cross Reference

Search

FT		Please take charge of search in sector between lat… and… , and long… and…	
FU		The search area of the aircraft is between lat… and… , and long… and…	
FV		Search by aircraft/helicopter will be discontinued because of unfavorable conditions.	
FW		You should search in the vicinity of lat… long…	
FX		Shall I search in the vicinity of lat… long…?	
FY		I am in the search area.	
	FY 1	Are you in the search area?	
		Aircraft is coming to participate in search. Expected arrive over the area of accident at time indicated.	**BP**
FZ		You should continue search according to instructions and until further notice.	
	FZ 1	I am continuing to search.	
	FZ 2	Are you continuing to search?	
	FZ 3	Do you want me to continue to search?	
GA		I cannot continue to search.	
GB		You should stop search and return to base or continue your voyage.	

* These signals are intended for use in connection with the Merchant Ship Search and Rescue Manual (MERSAR).

Results of Search

GC		Report results of search.	
	GC 1	Results of search negative. I am continuing to search.	
	GC 2	I have searched area of accident but have found no trace of derelict or survivors.	
	GC 3	I have noted patches of oil at likely position of accident.	
GD		Vessel/aircraft missing or being looked for has not been heard of since.	
	GD 1	Have you anything to report on vessels/aircraft missing or being looked for?	
	GD 2	Have you seen wreckage (or derelict)?	
GE		Vessel/aircraft has been located at lat… long…	
GF		I have found vessel/aircraft in distress in lat… long…	
GG		Vessel/aircraft was last reported at time indicated in lat… long… steering course…	
GH		I have sighted survival craft in lat… long… (or bearing… distance… from me).	

Code		*Meaning*	*Cross Reference*

Results of Search

GI		Survival craft are believed to be in the vicinity of lat… long…	
GJ		Wreckage is reported in lat… long…	
	GJ 1	Wreckage is reported in lat… long… No survivors appear to be in the vicinity.	
GK		Aircraft wreckage is found in lat… long…	
GL		I have located (or found) wreckage from the vessel/aircraft in distress (position to be indicated if necessary by lat… and long… or by bearing… from specified place and distance…).	

Rescue

GM		I cannot save my vessel.	
	GM 1	I cannot save my vessel; keep as close as possible.	
GN		You should take off persons.	
	GN 1	I wish some persons taken off. Skeleton crew will remain on board.	
	GN 2	I will take off persons.	
	GN 3	Can you take off persons?	
GO		I cannot take off persons.	
GP		You should proceed to the rescue of vessel (or ditched aircraft) in lat… long…	
GQ		I cannot proceed to the rescue owing to weather. You should do all you can.	
GR		Vessel coming to your rescue (or to the rescue of vessel or aircraft indicated) is steering course… , speed… knots.	
	GR 1	You should indicate course and speed of vessel coming to my rescue (or to the rescue of vessel or aircraft indicated).	
GS		I will attempt rescue with whip and breeches buoy.	
*GT		I will endeavor to connect with line throwing apparatus.	
	GT 1	Look out for rocket line.	
GU		It is not safe to fire a rocket.	
GV		You should endeavor to send me a line.	
	GV 1	Have you a line throwing apparatus?	
	GV 2	Can you connect with line throwing apparatus?	
	GV 3	I have not a line throwing apparatus.	
GW		Man overboard. Please take action to pick him up (position to be indicated if necessary).	
		Man overboard	**O**

* Reference is made to signals prescribed by the International Convention for the Safety of Life at Sea, 1974 (Regulation 16(c) Chapter V) in connection with the use of shore lifesaving apparatus.

Results of Rescue

Code	Meaning
GX	Report results of rescue.
GX 1	What have you (or rescue vessel/aircraft) picked up?
GY	I (or rescue vessel/aircraft) have picked up wreckage.
GZ	All persons saved.
GZ 1	All persons lost.
HA	I (or rescue vessel/aircraft) have rescued… (number) injured persons.
HB	I (or rescue vessel/aircraft) have rescued… (number) survivors.
HC	I (or rescue vessel/aircraft) have picked up… (number) bodies.
HD	Can I transfer rescued persons to you?

SURVIVORS

Code	Meaning
HF	I have located survivors in water, lat…long…(or bearing…from place indicated, distance…).
HG	I have located survivors in survival craft lat… long… (or bearing… from place indicated, distance…).
HJ	I have located survivors on drifting ice, lat… long…
HK	I have located bodies in lat… long… (or bearing… from place indicated, distance…).
HL	Survivors not yet located.
HL 1	I am still looking for survivors.
HL 2	Have you located survivors? If so, in what position?
HM	Survivors are in bad condition. Medical assistance is urgently required.
HM 1	Survivors are in bad condition.
HM 2	Survivors are in good condition.
HM 3	Condition of survivors not ascertained.
HM 4	What is condition of survivors?
HN	You should proceed to lat… long… to pick up survivors.
HO	Pick up survivors from drifting ice, lat… long…

CHAPTER 2.—GENERAL SIGNAL CODE

Code		Meaning	Cross Reference
	HO 1	Pick up survivors from sinking vessel/aircraft.	
HP		Survivors have not yet been picked up.	
	HP 1	Have survivors been picked up?	
		You should give immediate assistance to pick up survivors.	CN 1
HQ		Transfer survivors to my vessel (or vessel indicated).	
	HQ 1	Have you any survivors on board?	
HR		You should try to obtain from survivors all possible information.	
HT		You should indicate position of survivors by throwing pyrotechnic signals.	

CHAPTER 2

SECTION 2: CASUALTIES—DAMAGES

Meaning

COLLISION

Code		Meaning	Cross Reference
HV		Have you been in collision?	
HW		I have (or vessel indicated has) collided with surface craft.	
	HW 1	I have (or vessel indicated has) collided with light vessel.	
	HW 2	I have (or vessel indicated has) collided with submarine.	
	HW 3	I have (or vessel indicated has) collided with unknown vessel.	
	HW 4	I have (or vessel indicated has) collided with underwater object.	
	HW 5	I have (or vessel indicated has) collided with navigation buoy.	
	HW 6	I have (or vessel indicated has) collided with iceberg.	
	HW 7	I have (or vessel indicated has) collided with floating ice.	
HX		Have you received any damage in collision?	
	HX 1	I have received serious damage above the waterline.	
	HX 2	I have received serious damage below the waterline.	
	HX 3	I have received minor damage above the waterline.	
	HX 4	I have received minor damage below the waterline.	
HY		The vessel (name or identity signal) with which I have been in collision has sunk.	
	HY 1	The vessel (name or identity signal) with which I have been in collision has resumed her voyage.	
	HY 2	I do not know what has happened to the vessel with which I collided.	
	HY 3	Has the vessel with which you have been in collision resumed her voyage?	
	HY 4	What is the name (or identity signal) of the vessel with which you collided?	
	HY 5	What is the name (or identity signal) of vessel which collided with me? My name (or identity signal) is…	
	HY 6	Where is the vessel with which you collided?	
HZ		There has been a collision between vessels indicated (names or identity signals).	
		I urgently require a collision mat.	**KA**
		I have placed the collision mat. I can proceed without assistance.	**KA 1**
		Can you place the collision mat?	**KA 2**

DAMAGES—REPAIRS

Code	Meaning	Cross Reference
IA	I have received damage to stem.	

CHAPTER 2.—GENERAL SIGNAL CODE

Code		Meaning	Cross Reference
	IA 1	I have received damage to stern frame.	
	IA 2	I have received damage to side plate above water.	
	IA 3	I have received damage to side plate below water.	
	IA 4	I have received damage to bottom plate.	
	IA 5	I have received damage to boilerroom.	
	IA 6	I have received damage to engineroom.	
	IA 7	I have received damage to hatchways.	
	IA 8	I have received damage to steering gear.	
	IA 9	I have received damage to propellers.	
IB		What damage have you received?	
	IB 1	My vessel is seriously damaged.	
	IB 2	I have minor damage.	
	IB 3	I have not received any damage.	
	IB 4	The extent of the damage is still unknown.	
		Have you received any damage in collision?	HX
		I have received serious damage above the waterline	HX 1
		I have received serious damage below the waterline	HX 2
		I have received minor damage above the waterline	HX 3
		I have received minor damage below the waterline	HX 4
IC		Can damage be repaired at sea?	
	IC 1	Can damage be repaired at sea without assistance?	
	IC 2	How long will it take you to repair damage?	
ID		Damage can be repaired at sea.	
	ID 1	Damage can be repaired at sea without assistance.	
	ID 2	Damage has been repaired.	
IF		Damage cannot be repaired at sea.	
	IF 1	Damage cannot be repaired at sea without assistance.	
IG		Damage can be repaired in… (number) hours.	
IJ		I will try to proceed by my own means but I request you to keep in contact with me by… (Complements Table 1, Chapter 2, Section 10, Page 104).	
IK		I can proceed at… (number) knots.	

50

Code		Meaning	Cross Reference
IL		I can only proceed at slow speed.	
	IL 1	I can only proceed with one engine.	
	IL 2	I am unable to proceed under my own power.	
	IL 3	Are you in a condition to proceed?	
IM		I request to be escorted until further notice.	
		Propeller shaft is broken	**RO**
		My propeller is fouled by hawser or rope	**RO 1**
		I have lost my propeller	**RO 2**

DIVER—UNDERWATER OPERATIONS

Code		Meaning	Cross Reference
IN		I require a diver.	
	IN 1	I require a diver to clear propeller.	
	IN 2	I require a diver to examine bottom.	
	IN 3	I require a diver to place collision mat.	
	IN 4	I require a diver to clear my anchor.	
IO		I have no diver.	
IP		A diver will be sent as soon as possible (or at time indicated).	
IQ		Diver has been attacked by diver's disease and requires decompression chamber treatment.	
*IR		I am engaged in submarine survey work (underwater operations). Keep clear of me and go slow.	
		I have a diver down; keep well clear at slow speed	**A**

* The use of this signal does not relieve any vessel from compliance with the International Regulations for Preventing Collisions at Sea 1972.

FIRE—EXPLOSION

Fire

Code		Meaning	Cross Reference
IT		I am on fire.	
	IT 1	I am on fire and have dangerous cargo on board; keep well clear of me	**J**
	IT 2	Vessel (name or identity signal) is on fire.	
	IT 3	Are you on fire?	

Code		Meaning	Cross Reference

Fire

Code		Meaning	Cross Reference
IU		Vessel (name or identity signal) on fire is located at lat… long…	
		I require immediate assistance; I am on fire.	CB 6
IV		Where is the fire?	
	IV 1	I am on fire in the engineroom.	
	IV 2	I am on fire in the boilerroom.	
	IV 3	I am on fire in hold or cargo.	
	IV 4	I am on fire in passenger's or crew's quarters.	
	IV 5	Oil is on fire.	
IW		Fire is under control.	
IX		Fire is gaining.	
	IX 1	I cannot get the fire under control without assistance.	
	IX 2	Fire has not been extinguished.	
IY		I can get the fire under control without assistance.	
	IY 1	Can you get the fire under control without assistance?	
IZ		Fire has been extinguished.	
	IZ 1	I am flooding compartment to extinguish fire.	
	IZ 2	Is fire extinguished?	
JA		I require firefighting appliances.	
	JA 1	I require foam fire extinguishers.	
	JA 2	I require CO_2 fire extinguishers.	
	JA 3	I require tetrachloride fire extinguishers.	
	JA 4	I require material for foam fire extinguishers.	
	JA 5	I require material for CO_2 fire extinguishers.	
	JA 6	I require material for carbon tetrachloride fire extinguishers.	
	JA 7	I require water pumps.	

Explosion

Code		Meaning	Cross Reference
JB		There is danger of explosion.	
JC		There is no danger of explosion.	

Code		Meaning	Cross Reference

Explosion

	JC 1	Is there any danger of explosion?	
JD		Explosion has occurred in boiler.	
	JD 1	Explosion has occurred in tank.	
	JD 2	Explosion has occurred in cargo.	
	JD 3	Further explosions are possible.	
	JD 4	There is danger of toxic effects.	
JE		Have you any casualties owing to explosion?	
		An explosion was seen or heard (position or direction and time to be indicated)	**EK 1**

GROUNDING—BEACHING—REFLOATING

Grounding

JF		I am (or vessel indicated is) aground in lat... long... (also the following complements, if necessary):	
	0	On rocky bottom.	
	1	On soft bottom.	
	2	Forward.	
	3	Amidship.	
	4	Aft.	
	5	At high water forward.	
	6	At high water amidship.	
	7	At high water aft.	
	8	Full length of vessel.	
	9	Full length of vessel at high water.	
JG		I am aground; I am in dangerous situation.	
JH		I am aground; I am not in danger.	
		I require immediate assistance; I am aground	**CB 4**
		Vessel aground in lat... long... require assistance	**CI**
JI		Are you aground?	
	JI 1	What was your draft when you went aground?	

Code		Meaning	Cross Reference

Grounding

	JI 2	On what kind of ground have you gone aground?	
	JI 3	At what state of tide did you go aground?	
	JI 4	What part of your vessel is aground?	
JJ		My maximum draft when I went aground was… (number) feet or meters.	
JK		The tide was high water when the vessel went aground.	
	JK 1	The tide was half water when the vessel went aground.	
	JK 2	The tide was low water when the vessel went aground.	
JL		You are running the risk of going aground.	
	JL 1	You are running the risk of going aground; do not approach me from the starboard side.	
	JL 2	You are running the risk of going aground; do not approach me from the port side.	
	JL 3	You are running the risk of going aground; do not approach me from forward.	
	JL 4	You are running the risk of going aground; do not approach me from aft.	
JM		You are running the risk of going aground at low water.	

Beaching

JN		You should beach the vessel in lat… long…	
	JN 1	You should beach the vessel where flag is waved or light is shown.	
	JN 2	I must beach the vessel.	

Refloating

JO		I am afloat.	
	JO 1	I am afloat forward.	
	JO 2	I am afloat aft.	
	JO 3	I may be got afloat if prompt assistance is given.	
	JO 4	Are you (or vessel indicated) still afloat?	
	JO 5	When do you expect to be afloat?	
JP		I am jettisoning to refloat (the following complements should be used if required):	
	1	Cargo.	
	2	Bunkers.	
	3	Everything movable forward.	

Code		Meaning	Cross Reference

Refloating

	4	Everything movable aft.	
JQ		I cannot refloat without jettisoning (the following complements should be used if required):	
	1	Cargo.	
	2	Bunkers.	
	3	Everything movable forward.	
	4	Everything movable aft.	
JR		I expect (or vessel indicated expects) to refloat.	
	JR 1	I expect (or vessel indicated expects) to refloat at time indicated.	
	JR 2	I expect (or vessel indicated expects) to refloat in daylight.	
	JR 3	I expect (or vessel indicated expects) to refloat when tide rises.	
	JR 4	I expect (or vessel indicated expects) to refloat when visibility improves.	
	JR 5	I expect (or vessel indicated expects) to refloat when weather moderates.	
	JR 6	I expect (or vessel indicated expects) to refloat when draft is lightened.	
	JR 7	I expect (or vessel indicated expects) to refloat when tugs arrive.	
JS		Is it likely that you (or vessel indicated) will refloat?	
	JS 1	Is it likely that you (or vessel indicated) will refloat at time indicated?	
	JS 2	Is it likely that you (or vessel indicated) will refloat in daylight?	
	JS 3	Is it likely that you (or vessel indicated) will refloat when tide rises?	
	JS 4	Is it likely that you (or vessel indicated) will refloat when visibility improves?	
	JS 5	Is it likely that you (or vessel indicated) will refloat when weather moderates?	
	JS 6	Is it likely that you (or vessel indicated) will refloat when draft is lightened?	
	JS 7	Is it likely that you (or vessel indicated) will refloat when tugs arrive?	
JT		I can refloat if an anchor is laid out for me.	
	JT 1	I may refloat without assistance.	
	JT 2	Will you assist me to refloat?	
JU		I cannot be refloated by any means now available.	
JV		Will you escort me to lat... long... after refloating?	

Code		Meaning	Cross Reference

LEAK

Code		Meaning	Cross Reference
JW		I have sprung a leak.	
	JW 1	Leak is dangerous.	
	JW 2	Leak is causing dangerous heel.	
	JW 3	Leak is beyond the capacity of my pumps.	
		I require immediate assistance; I have sprung a leak	**CB 7**
JX		Leak is gaining rapidly.	
	JX 1	I cannot stop the leak.	
JY		Leak can be controlled, if it does not get any worse.	
	JY 1	I require additional pumping facilities to control the leak.	
	JY 2	Leak is under control.	
	JY 3	Leak has been stopped.	
JZ		Have you sprung a leak?	
	JZ 1	Can you stop the leak?	
	JZ 2	Is the leak dangerous?	
KA		I urgently require a collision mat.	
	KA 1	I have placed the collision mat. I can proceed without assistance.	
	KA 2	Can you place the collision mat?	
KB		I have… (number) feet or meters of water in the hold.	
KC		My hold(s) is (are) flooded.	
	KC 1	How many compartments are flooded?	
KD		There are… (number) compartments flooded.	
KE		The watertight bulkheads are standing up well to the pressure of the water.	
	KE 1	I need timber to support bulkheads.	

TOWING—TUGS

Tug

Code	Meaning	Cross Reference
KF	I require a tug (or… (number) tugs).	
	I require a tug	**Z**

Code		Meaning	Cross Reference

Tug

KG		Do you require a tug(s)?
	KG 1	I do not require tug(s).
KH		Tug(s) is (are) coming to you. Expect to arrive at time indicated.
	KH 1	Tug with pilot is coming to you.
	KH 2	You should wait for tugs.
KI		There are no tugs available.
	KI 1	Tugs cannot proceed out.

Towing—Taking in Tow

KJ		I am towing a submerged object.
	KJ 1	I am towing a float.
	KJ 2	I am towing a target.
KK		Towing is impossible under present weather conditions.
	KK 1	Towing is very difficult.
	KK 2	I cannot connect at present but will attempt when conditions improve.
	KK 3	I cannot connect tonight. I will try in daylight.
	KK 4	Can you assist with your engines?
KL		I am obliged to stop towing temporarily.
	KL 1	You should stop towing temporarily.
KM		I can take you (or vessel indicated) in tow.
	KM 1	Shall I take you in tow?
KN		I cannot take you (or vessel indicated) in tow.
	KN 1	I cannot take you (or vessel indicated) in tow but I will report you and ask for immediate assistance.
	KN 2	I cannot take you (or vessel indicated) in tow but can take off persons.
KO		You should endeavor to take vessel (name or identity signal) in tow.
	KO 1	You should report whether you have taken vessel (name or identity signal) in tow.
	KO 2	Can you take me (or vessel indicated) in tow?
KP		You should tow me to the nearest port or anchorage (or place indicated).
	KP 1	I will tow you to the nearest port or anchorage (or place indicated).

Towing—Taking in Tow

Code	Meaning
KP 2	I must get shelter or anchorage as soon as possible.
KQ	Prepare to be taken in tow.
KQ 1	I am ready to be taken in tow.
KQ 2	Prepare to tow me (or vessel indicated).
KQ 3	I am ready to tow you.
KQ 4	Prepare to resume towing.
KQ 5	I am ready to resume towing.
KR	All is ready for towing.
KR 1	I am commencing to tow.
KR 2	You should commence towing.
KR 3	Is all ready for towing?

Towing Line—Cable—Hawser

Code	Meaning
KS	You should send a line over.
KS 1	I have taken the line.
KT	You should send me a towing hawser.
KT 1	I am sending towing hawser.
KU	I cannot send towing hawser.
KU 1	I have no, or no other, hawser.
KU 2	I have no wire hawser.
KU 3	Have you a hawser?
KV	I intend to use my towing hawser/cable.
KV 1	I intend to use your towing hawser/cable.
KW	You should have towing hawser/cable ready.
KW 1	Towing hawser/cable is ready.
KW 2	You should have another hawser ready.
KW 3	You should have spare towing hawser/cable ready.
KW 4	Spare towing hawser/cable is ready.
KW 5	You should have wire hawser ready.

Code		Meaning	Cross Reference

Towing Line—Cable—Hawser

	KW 6	Wire hawser is ready.	
KX		You should be ready to receive the towing hawser.	
	KX 1	I am ready to receive the towing hawser.	
	KX 2	You should come closer to receive towing hawser.	
	KX 3	I am coming closer to receive towing hawser.	
	KX 4	I have received towing hawser.	
KY		Length of tow is… (number) fathoms.	
KZ		You should shorten the towing hawser (or shorten distance between vessels).	
	KZ 1	I am shortening towing hawser (or I am shortening distance between vessels).	
	KZ 2	You should haul in the hawser.	
	KZ 3	I am hauling in the hawser.	
	KZ 4	You should haul in the slack.	
	KZ 5	I am hauling in the slack.	
LA		Towing hawser/cable has parted.	
	LA 1	Towing hawser/cable is in danger of parting.	
	LA 2	Towing hawser/cable is damaged.	
	LA 3	You should reinforce the hawsers.	
	LA 4	I am reinforcing the hawsers.	

Make Fast—Veer

LB		You should make towing hawser fast to your chain cable.	
	LB 1	Towing hawser is fast to chain cable.	
	LB 2	You should make towing hawser fast to wire.	
	LB 3	Towing hawser is fast to wire.	
	LB 4	My towing hawser is fast.	
	LB 5	Is your towing hawser fast?	
LC		You should make fast astern and steer me.	
LD		You should veer your hawser/cable… (number) fathoms.	
LE		I am about to veer my hawser/cable.	

Code	Meaning	Cross Reference

Make Fast—Veer

Code	Meaning
LE 1	I am veering my hawser/cable.
LE 2	I have veered my hawser/cable.
LE 3	I shall veer cable attached to hawser.
LE 4	How much cable should I veer?
LF	You should stop veering your hawser/cable.
LF 1	I cannot veer any more hawser/cable.

Cast Off

Code	Meaning
LG	You should prepare to cast off towing hawser(s).
LG 1	I am preparing to cast off towing hawser(s).
LG 2	I am ready to cast off towing hawser(s).
LG 3	You should cast off starboard towing hawser.
LG 4	I have cast off starboard towing hawser.
LG 5	You should cast off port towing hawser.
LG 6	I have cast off port towing hawser.
LG 7	You should cast off hawser(s).
LG 8	I must cast off towing hawser(s).

Engine Maneuvers

Meaning	Cross Reference
I am going ahead	**QD**
My engines are going ahead	**QD 1**
I will keep going ahead.	**QD 2**
I will go ahead	**QD 3**
I will go ahead dead slow	**QD 4**
I have headway	**QE**
I cannot go ahead	**QF**
You should go ahead.	**QG**
You should go slow ahead	**QG 1**
You should go full speed ahead	**QG 2**
You should keep going ahead.	**QG 3**

Code	Meaning	Cross Reference

Engine Maneuvers

	Meaning	Cross Reference
	You should keep your engines going ahead	**QG 4**
	You should not go ahead any more	**QH**
	I am going astern	**QI**
	My engines are going astern	**QI 1**
	I will keep going astern	**QI 2**
	I will go astern	**QI 3**
	I will go astern dead slow.	**QI 4**
	I have sternway	**QJ**
	I cannot go astern.	**QK**
	You should go astern	**QL**
	You should go slow astern	**QL 1**
	You should go full speed astern	**QL 2**
	You should keep going astern	**QL 3**
	You should keep your engines going astern.	**QL 4**
	You should not go astern any more	**QM**
	You should stop your engines immediately	**RL**
	You should stop your engines	**RL 1**
	My engines are stopped	**RM**
	I am stopping my engines	**RM 1**

Code		Meaning
LH		Maximum speed in tow is... (number) knots.
LI		I am increasing speed.
	LI 1	Increase speed.
LJ		I am reducing speed.
	LJ 1	Reduce speed.

CHAPTER 2

SECTION 3: AIDS TO NAVIGATION—NAVIGATION—HYDROGRAPHY

Code	*Meaning*	*Cross Reference*

AIDS TO NAVIGATION

Buoys—Beacons

LK	Buoy (or beacon) has been established in lat… long…	
LL	Buoy (or beacon) in lat… long… has been removed.	
	You should steer directly for the buoy (or object indicated)	**PL**
	You should keep buoy (or object indicated) on your starboard side	**PL 1**
	You should keep buoy (or object indicated) on your port side.	**PL 2**
	You can pass the buoy (or object indicated) on either side	**PL 3**
LM	Radiobeacon indicated is out of action.	

Lights—Lightvessels

LN		Light (name follows) has been extinguished.	
	LN 1	All lights are out along this coast (or the coast of…).	
LO		I am not in my correct position (to be used by a lightvessel).	
	LO 1	Lightvessel (name follows) is out of position.	
	LO 2	Lightvessel (name follows) has been removed from her station.	
		Lightvessel (or lighthouse) indicated requires assistance	**CH 1**

BAR

LP		There is not less than… (number) feet or meters of water over the bar.
LQ		There will be… (number) feet or meters of water over the bar at time indicated.
LR		Bar is not dangerous.
	LR 1	What is the depth of water over the bar?
	LR 2	Can I cross the bar?
LS		Bar is dangerous.

BEARINGS

LT	Your bearing from me [or from… (name or identity signal)] is… (at time indicated).	
LU	The bearing of… (name or identity signal) from… (name or identity signal) is… (at time indicated).	
LV	Let me know my bearings from you. I will flash searchlight.	

Code		Meaning	Cross Reference
	LV 1	What is my bearing from you [or from… (name or identity signal)]?	
	LV 2	What is the bearing of… (name or identity signal) from… (name or identity signal)?	
		Your magnetic bearing from me (or from vessel or position indicated) is… (at time indicated)	**BZ**
		What is my magnetic bearing from you (or from vessel or position indicated)?.	**CA**
LW		I receive your transmission on bearing…	
	LW 1	Can you take bearings from my radio signals?	
		Your position according to bearings taken by radio direction finder stations which I control is lat… long… (at time indicated)	**EZ**
		Will you give me my position according to bearings taken by radio direction finder stations which you control?.	**EZ 1**
		Bearing and distance by radar of vessel (or object) indicated is bearing… , distance… miles.	**OM**
		What is the bearing and distance by radar of vessel (or object) indicated?.	**OM 1**

CANAL—CHANNEL—FAIRWAY

Canal

LX		The canal is clear.
	LX 1	The canal will be clear at time indicated.
	LX 2	You can enter the canal at time indicated.
	LX 3	Is the canal clear?
	LX 4	When can I enter the canal?
LY		The canal is not clear.
LZ		The channel/fairway is navigable.
	LZ 1	I intend to pass through the channel/fairway.
	LZ 2	Is the channel/fairway navigable?
	LZ 3	What is the state of the channel/fairway?
	LZ 4	What is the least depth of water in the channel/fairway?
MA		The least depth of water in the channel/fairway is… (number feet or meters).
MB		You should keep in the center of the channel/fairway.
	MB 1	You should keep on the starboard side of the channel/fairway.

Code		Meaning	Cross Reference

Canal

	MB 2	You should keep on the port side of the channel/fairway.	
	MB 3	You should leave the channel/fairway free.	
MC		There is an uncharted obstruction in the channel/fairway. You should proceed with caution.	
	MC 1	The channel/fairway is not navigable.	
	MC 2	The (—) lane of the traffic separation scheme is not navigable. (The direction of the traffic flow is to be indicated.)	

COURSE

	MD	My course is....	
	MD 1	What is your course?	
		My present position, course, and speed are lat... long... ,... , knots.....................	EV
		What are your present position, course, and speed?...............................	EV 1
		Will vessels in my immediate vicinity (or in the vicinity of lat... long...) please indicate position, course, and speed...	FB
		Vessel coming to your rescue (or to the rescue of vessel or aircraft indicated) is steering course... , speed... knots	GR
		You should indicate course and speed of vessel coming to my rescue (or to the rescue of vessel or aircraft indicated)...	GR 1
ME		The course to place (name follows) is...	
	ME 1	What is the course to place (name follows)?	
		The magnetic course for the helicopter to regain its base is...........................	BX
MF		Course to reach me is...	
	MF 1	What is the course to reach you?	
		The magnetic course for you to steer towards me (or vessel or position indicated) is... (at time indicated)...	BW
		Will you indicate the magnetic course for me to steer towards you (or vessel or position indicated)?	BY
MG		You should steer course...	
	MG 1	What course should I steer?	
		You should maintain your present course..	PI
		I am maintaining my present course..	PI 1

Code	Meaning	Cross Reference
	I cannot maintain my present course	**PJ**
MH	You should alter course to… (at time indicated).	
MI	I am altering course to…	
	I am altering my course to starboard	**E**
	I am altering my course to port	**I**
	You should alter your course, if possible, appreciably to starboard to facilitate location by radar	**OJ 2**
	You should alter your course, if possible, appreciably to port to facilitate location by radar . .	**OJ 3**

DANGERS TO NAVIGATION—WARNINGS

Derelict—Wreck—Shoal

MJ	Derelict dangerous to navigation reported in lat… long… (or Complements Table 3, Chapter 2, Section 10, Page 104).	
MK	I have seen derelict (in lat… long… at time indicated).	
MK 1	Have you seen derelict (or wreckage)?	
ML	Derelict is drifting near lat… long… (or bearing… from place indicated, distance…).	
ML 1	Hull of derelict is awash.	
ML 2	Hull of derelict is well out of the water.	
MM	There is a wreck in lat… long…	
MM 1	Wreck is buoyed.	
MM 2	Wreck is awash.	
MN	Wreck (in lat… long…) is not buoyed.	
MO	I have struck a shoal or submerged object (lat… long…).	
MP	I am in shallow water. Please direct me how to navigate.	

Radiation Danger

MQ	There is risk of contamination due to excessive release of radioactive material in this area (or in area around lat… long…). Keep radio watch. Relay the message to vessels in your vicinity.	
MQ 1	The radioactive material is airborne.	
MQ 2	The radioactive material is waterborne.	
MR	There is no, or no more, risk of contamination due to excessive release of radioactive material in this area (or in area around lat… long…).	

Code		Meaning	Cross Reference

Radiation Danger

MR 1 — Is there risk of contamination due to excessive release of radioactive material in this area (or in area around lat… long…)?

MS — My vessel is a dangerous source of radiation.

MS 1 — My vessel is a dangerous source of radiation; you may approach from my starboard side.

MS 2 — My vessel is a dangerous source of radiation; you may approach from my port side.

MS 3 — My vessel is a dangerous source of radiation; you may approach from forward.

MS 4 — My vessel is a dangerous source of radiation; you may approach from aft.

MT — My vessel is a dangerous source of radiation. You may approach from… (Complements Table 3, Chapter 2, Section 10, Page 104).

MU — My vessel is a dangerous source of radiation. Do not approach within… (number) cables.

I am abandoning my vessel which has suffered a nuclear accident and is a possible source of radiation danger . . . . **AD**

I have had a serious nuclear accident and you should approach with caution . . . . **AJ**

I have had a nuclear accident on board . . . . **AK**

A vessel which has suffered a nuclear accident is in distress in lat… long… . . . . **EC**

MV — My vessel is releasing radioactive material and presents a hazard.

MW — My vessel is releasing radioactive material and presents a hazard. Do not approach within… (number) cables.

MX — The radioactive material is airborne. Do not approach from leeward.

Warnings

MY — It is dangerous to stop.

MY 1 — It is dangerous to remain in present position.

MY 2 — It is dangerous to proceed on present course.

MY 3 — It is dangerous to proceed until weather permits.

MY 4 — It is dangerous to alter course to starboard.

MY 5 — It is dangerous to alter course to port.

MY 6 — It is dangerous to approach close to my vessel.

MY 7 — It is dangerous to let go an anchor or use bottom trawl.

MY 8 — It is dangerous to jettison inflammable oil.

It is not safe to fire a rocket . . . . **GU**

SECTION 3.—AIDS TO NAVIGATION—NAVIGATION—HYDROGRAPHY

Code		Meaning	Cross Reference

Warnings

MZ		Navigation is dangerous in the area around lat… long…	
	MZ 1	I consider you are carrying out a dangerous navigational practice and I intend to report you.	
		Navigation is dangerous in the area around lat… long… owing to iceberg(s)	**VZ**
		Navigation is dangerous in the area around lat… long… owing to floating ice	**VZ 1**
		Navigation is dangerous in the area around lat… long… owing to pack ice.	**VZ 2**
NA		Navigation is closed.	
	NA 1	Navigation is possible only with tug assistance.	
	NA 2	Navigation is possible only with pilot assistance.	
	NA 3	Navigation is prohibited within 500m of this platform.	
	NA 4	Navigation is prohibited within 500m of the platform bearing (—) from me.	
	NA 5	You have been detected navigating within a 500m Safety Zone (about the platform bearing (—) from me) and will be reported.	
	NA 6	Anchors with buoys extend up to one mile from this vessel/rig. You should keep clear.	
		You should navigate with caution. Small fishing boats are within… (number) miles of me.	**TH**
		You should navigate with caution. You are drifting towards my set of nets.	**TI**
		You should navigate with caution. There are nets with a buoy in this area.	**TJ**
NB		There is fishing gear in the direction you are heading (or in direction indicated—Complements Table 3, Chapter 2, Section 10, Page 104).	
NC		I am in distress and require immediate assistance.	
ND		Tsunami (phenomenal wave) is expected. You should take appropriate precautions.	
		Tropical storm (cyclone, hurricane, typhoon) is approaching. You should take appropriate precautions.	**VL**
NE		You should proceed with great caution.	
	NE 1	You should proceed with great caution; the coast is dangerous.	
	NE 2	You should proceed with great caution; submarines are exercising in this area.	
	NE 3	You should proceed with great caution; there is a boom across.	
	NE 4	You should proceed with great caution; keep clear of firing range.	
	NE 5	You should proceed with great caution; hostile vessel sighted (in lat… long…).	
	NE 6	You should proceed with great caution; hostile submarine sighted (in lat… long…).	

Code		Meaning	Cross Reference

Warnings

	NE 7	You should proceed with great caution; hostile aircraft sighted (in lat... long...).	
		There is an uncharted obstruction in the channel/fairway. You should proceed with caution	**MC**
		You should change your anchorage/berth. It is not safe.	**RE**
		All vessels should proceed to sea as soon as possible owing to danger in port	**UL**
NF		You are running into danger	**U**
	NF 1	You are running into danger; there is a radiation hazard.	
NG		You are in a dangerous position.	
	NG 1	You are in a dangerous position; there is a radiation hazard.	
NH		You are clear of all danger.	
	NH 1	Are you clear of all danger?	
NI		I have (or vessel indicated has) a list of... (number) degrees to starboard.	
NJ		I have (or vessel indicated has) a list of... (number) degrees to port.	

DEPTH—DRAFT

Depth

NK		There is not sufficient depth of water.	
NL		There is sufficient depth of water.	
	NL 1	Is there sufficient depth of water?	
		The least depth of water in the channel/fairway is... (number feet or meters)	**MA**
		What is the least depth of water in the channel/fairway.	**LZ 4**
		There is not less than... (number feet or meters) of water over the bar.	**LP**
		What is the depth of water over the bar?	**LR 1**
		There will be... (number feet or meters) of water over the bar at time indicated.	**LQ**
		The depth at high water here (or in place indicated) is... (number feet or meters).	**QA**
		The depth at low water here (or in place indicated) is... (number feet or meters).	**QB**
		What is the depth at high and low water here (or in place indicated)?.	**PW 2**
NM		You should report the depth around your vessel.	
NN		I am in... (number feet or meters) of water.	

Code		*Meaning*	*Cross Reference*

Depth

*NO		Negative—"No" or "The significance of the previous group should be read in the negative".	
NP		The depth of water at the bow is… (number feet or meters).	
NQ		The depth of water at the stern is… (number feet or meters).	
NR		The depth of water along the starboard side is… (number feet or meters).	
NS		The depth of water along the port side is… (number feet or meters).	

* Procedural signal.

Draft

NT		What is your draft?	
	NT 1	What is your light draft?	
	NT 2	What is your ballast draft?	
	NT 3	What is your loaded draft?	
	NT 4	What is your summer draft?	
	NT 5	What is your winter draft?	
	NT 6	What is your maximum draft?	
	NT 7	What is your least draft?	
	NT 8	What is your draft forward?	
	NT 9	What is your draft aft?	
NU		My draft is… (number feet or meters).	
NV		My light draft is… (number feet or meters).	
NW		My ballast draft is… (number feet or meters).	
NX		My loaded draft is… (number feet or meters).	
NY		My summer draft is… (number feet or meters).	
NZ		My winter draft is… (number feet or meters).	
OA		My maximum draft is… (number feet or meters).	
OB		My least draft is… (number feet or meters).	
OC		My draft forward is… (number feet or meters).	
OD		My draft aft is… (number feet or meters).	
		My maximum draft when I went aground was (number feet or meters)	**JJ**

Code	Meaning	Cross Reference

Draft

	What was your draft when you went aground?	**JI 1**
OE	Your draft must not exceed… (number feet or meters).	
OF	I could lighten to… (number feet or meters) draft.	
OG	To what draft could you lighten?	

ELECTRONIC NAVIGATION

Radar

OH	You should switch on your radar and keep radar watch.	
OH 1	The restrictions on the use of radar are lifted.	
OH 2	Does my radar cause interference?	
OI	I have no radar.	
OI 1	Are you equipped with radar?	
OI 2	Is your radar in operation?	
OJ	I have located you on my radar bearing… , distance… miles.	
OJ 1	I cannot locate you on my radar.	
OJ 2	You should alter your course, if possible, appreciably to starboard to facilitate location by radar.	
OJ 3	You should alter your course, if possible, appreciably to port to facilitate location by radar.	
OJ 4	Can you locate me by radar?	
	My position is ascertained by radar	**EW 4**
***OK**	Acknowledging a correct repetition or "It is correct".	
OL	Is radar pilotage being effected in this port (or port indicated)?	
OM	Bearing and distance by radar of vessel (or object) indicated, is bearing… , distance… miles.	
OM 1	What is the bearing and distance by radar of vessel (or object) indicated?	
ON	I have an echo on my radar on bearing… , distance… miles.	

*   Procedural signal.

Radio Direction Finder

OO	My radio direction finder is inoperative.	

Code		Meaning	Cross Reference

Radio Direction Finder

OP		I have requested... (name or identity signal) to send two dashes of ten seconds each or the carrier of his transmitter followed by his call sign.	
	OP 1	Will you request... (name or identity signal) to send two dashes of ten seconds each or the carrier of his transmitter followed by his call sign?	
	OP 2	Will you send two dashes of ten seconds each, or the carrier of your transmitter, followed by your call sign?	
		Your position according to bearings taken by radio direction finder stations which I control is lat... long... (at time indicated)...	**EZ**
		Will you give me my position according to bearings taken by radio direction finder stations which you control?..	**EZ 1**
OQ		I am calibrating radio direction finder or adjusting compasses.	

Electronic Position-Fixing System

| | | My position is ascertained by electronic position fixing system....................... | **EW 5** |

MINES—MINESWEEPING

OR		I have struck a mine.	
		I have a mine in my sweep (or net)...	**TO**
OS		There is danger from mines in this area (or area indicated).	
	OS 1	You should keep a lookout for mines.	
	OS 2	You are out of the dangerous zone.	
	OS 3	Am I out of the dangerous zone?	
	OS 4	Are you out of the dangerous zone?	
	OS 5	Is there any danger from mines in this area (or area indicated)?	
OT		Mine has been sighted in lat... long... (or in direction indicated—Complements Table 3, Chapter 2, Section 10, Page 104).	
OU		Mine(s) has (have) been reported in the vicinity (or in approximate position lat... , long...).	
OV		Mine(s) is (are) believed to be bearing... from me, distance... miles.	
OW		There is a minefield ahead of you. You should stop your vessel and wait for instructions.	
	OW 1	There is a minefield along the coast. You should not approach too close.	
OX		The approximate direction of the minefield is bearing... from me.	
OY		Port is mined.	

Code		*Meaning*	*Cross Reference*
	OY 1	Entrance is mined.	
	OY 2	Fairway is mined.	
	OY 3	Are there mines in the port entrance or fairway?	
OZ		The width of the swept channel is… (number feet or meters).	
PA		I will indicate the swept channel. You should follow in my wake.	
	PA 1	You should keep carefully to the swept channel.	
	PA 2	The swept channel is marked by buoys.	
	PA 3	I do not see the buoys marking the swept channel.	
	PA 4	Do you know the swept channel?	
***PB**		You should keep clear of me; I am engaged in minesweeping operations.	
	PB 1	You should keep clear of me; I am exploding a floating mine.	
PC		I have destroyed the drifting mine(s).	
	PC 1	I cannot destroy the drifting mine(s).	

* The use of this signal does not relieve any vessel from complying with the International Regulations for Preventing Collisions at Sea 1972.

NAVIGATION LIGHTS—SEARCHLIGHT

PD		Your navigation light(s) is (are) not visible.	
	PD 1	My navigation lights are not functioning.	
PE		You should extinguish all the lights except the navigation lights.	
PG		I do not see any light.	
	PG 1	You should hoist a light.	
	PG 2	I am dazzled by your searchlight. Extinguish it or lift it.	
		You should train your searchlight nearly vertical on a cloud, intermittently if possible, and, if my aircraft is seen, deflect the beam upwind and on the water to facilitate my landing	**AX**
		Shall I train my searchlight nearly vertical on a cloud, intermittently if possible, and, if your aircraft is seen, deflect the beam upwind and on the water to facilitate your landing?	**AX 1**

Code		Meaning	Cross Reference

NAVIGATING AND STEERING INSTRUCTIONS

(See also Pilot in Chapter 2, Section 5, Page 87.)

PH   You should steer as indicated.

    PH 1   You should steer towards me.

    PH 2   I am steering towards you.

    PH 3   You should steer more to starboard.

    PH 4   I am steering more to starboard.

    PH 5   You should steer more to port.

    PH 6   I am steering more to port.

PI   You should maintain your present course.

    PI 1   I am maintaining my present course.

    PI 2   Shall I maintain my present course?

PJ   I cannot maintain my present course.

    You should make fast astern and steer me . . . .   **LC**

PK   I cannot steer without assistance.

PL   You should steer directly for the buoy (or object indicated).

    PL 1   You should keep buoy (or object indicated) on your starboard side.

    PL 2   You should keep buoy (or object indicated) on your port side.

    PL 3   You can pass the buoy (or object indicated) on either side.

PM   You should follow in my wake (or wake of vessel indicated).

    PM 1   You should go ahead and lead the course.

PN   You should keep to leeward of me (or vessel indicated).

    PN 1   You should keep to windward of me (or vessel indicated).

    PN 2   You should keep on my starboard side (or starboard side of vessel indicated).

    PN 3   You should keep on my port side (or port side of vessel indicated).

PO   You should pass ahead of me (or vessel indicated).

    PO 1   I will pass ahead of you (or vessel indicated).

    PO 2   You should pass astern of me (or vessel indicated).

    PO 3   I will pass astern of you (or vessel indicated).

    PO 4   You should pass to leeward of me (or vessel indicated).

Code		Meaning	Cross Reference
	PO 5	I will pass to leeward of you (or vessel indicated).	
	PO 6	You should pass to windward of me (or vessel indicated).	
	PO 7	I will pass to windward of you (or vessel indicated).	
	PO 8	You should come under my stern.	
PP		Keep well clear of me.	
	PP 1	Do not overtake me.	
	PP 2	Do not pass ahead of me.	
	PP 3	Do not pass astern of me.	
	PP 4	Do not pass on my starboard side.	
	PP 5	Do not pass on my port side.	
	PP 6	Do not pass too close to me.	
	PP 7	You should give way to me.	
PQ		You should keep closer in to the coast.	
	PQ 1	You should keep further away from the coast.	
	PQ 2	You should follow the coast at a safe distance.	
	PQ 3	How far out from the coast?	
PR		You should keep closer to me (or vessel indicated).	
	PR 1	You should come as near as possible.	
	PR 2	You should keep within visual signal distance from me (or vessel indicated).	
	PR 3	You should come within hailing distance from me (or vessel indicated).	
PS		You should not come any closer.	
	PS 1	You should keep away from me (or vessel indicated).	
		I am calibrating radio direction finder or adjusting compasses	OQ

TIDE

Code		Meaning	Cross Reference
PT		What is the state of the tide?	
	PT 1	The tide is rising.	
	PT 2	The tide is falling.	
	PT 3	The tide is slack.	
PU		The tide begins to rise at time indicated.	

SECTION 3.—AIDS TO NAVIGATION—NAVIGATION—HYDROGRAPHY

Code		Meaning	Cross Reference
	PU 1	When does the tide begin to rise?	
PV		The tide begins to fall at time indicated.	
	PV 1	When does the tide begin to fall?	
PW		What is the rise and fall of the tide?	
	PW 1	What is the set and drift of the tide?	
	PW 2	What is the depth at high and low water here (or in place indicated)?	
PX		The rise and fall of the tide is… (number feet or meters).	
PY		The set of the tide is… degrees.	
PZ		The drift of the tide is… knots.	
QA		The depth at high water here (or in place indicated) is… (number feet or meters).	
QB		The depth at low water here (or in place indicated) is… (number feet or meters).	
		The tide was high water when the vessel went aground	**JK**
		The tide was half water when the vessel went aground	**JK 1**
		The tide was low water when the vessel went aground	**JK 2**
		At what state of tide did you go aground?	**JI 3**
QC		You should wait until high water.	
	QC 1	You should wait until low water.	

CHAPTER 2

SECTION 4: MANEUVERS

AHEAD—ASTERN

Ahead—Headway

QD		I am going ahead.
	QD 1	My engines are going ahead.
	QD 2	I will keep going ahead.
	QD 3	I will go ahead.
	QD 4	I will go ahead dead slow.
QE		I have headway.
QF		I cannot go ahead.
QG		You should go ahead.
	QG 1	You should go slow ahead.
	QG 2	You should go full speed ahead.
	QG 3	You should keep going ahead.
	QG 4	You should keep your engines going ahead.
QH		You should not go ahead any more.

Astern—Sternway

QI		I am going astern.	
	QI 1	My engines are going astern	S
	QI 2	I will keep going astern.	
	QI 3	I will go astern.	
	QI 4	I will go astern dead slow.	
QJ		I have sternway.	
QK		I cannot go astern.	
QL		You should go astern.	
	QL 1	You should go slow astern.	
	QL 2	You should go full speed astern.	
	QL 3	You should keep going astern.	
	QL 4	You should keep your engines going astern.	
QM		You should not go astern any more.	

Code		*Meaning*	*Cross Reference*

ALONGSIDE

QN		You should come alongside my starboard side.	
	QN 1	You should come alongside my port side.	
	QN 2	You should drop an anchor before coming alongside.	
QO		You should not come alongside.	
QP		I will come alongside.	
	QP 1	I will try to come alongside.	
QQ		I require health clearance. (See Pratique Messages in Chapter 2, Section 9, Page 103.)	
QR		I cannot come alongside.	
	QR 1	Can I come alongside?	

TO ANCHOR—ANCHOR(S)—ANCHORAGE

To Anchor

QS		You should anchor at time indicated.	
	QS 1	You should anchor (position to be indicated if necessary).	
	QS 2	You should anchor to await tug.	
	QS 3	You should anchor with both anchors.	
	QS 4	You should anchor as convenient.	
	QS 5	Are you going to anchor?	
		You should heave to or anchor until pilot arrives.	**UB**
QT		You should not anchor. You are going to foul my anchor.	
QU		Anchoring is prohibited.	
QV		I am anchoring in position indicated.	
	QV 1	I have anchored with both anchors.	
QW		I shall not anchor.	
	QW 1	I cannot anchor.	
QX		I request permission to anchor.	
	QX 1	You have permission to anchor.	
QY		I wish to anchor at once.	

Code	Meaning	Cross Reference

To Anchor

QY 1   Where shall I anchor?

Anchor(s)

QZ   You should have your anchors ready for letting go.

    QZ 1   You should let go another anchor.

RA   My anchor is foul.

    RA 1   I have picked up telegraph cable with my anchor.

RB   I am dragging my anchor . . . .   **Y**

    RB 1   You appear to be dragging your anchor.

    RB 2   Where you have anchored (or intend to anchor) you are likely to drag.

RC   I am (or vessel indicated is) breaking adrift.

    RC 1   I have broken adrift.

RD   You should weigh (cut or slip) anchor immediately.

    RD 1   You should weigh anchor at time indicated.

    RD 2   I am unable to weigh my anchor.

Anchorage

RE   You should change your anchorage/berth. It is not safe.

RF   Will you lead me into a safe anchorage?

    You should tow me to the nearest port or anchorage (or place indicated) . . . .   **KP**

    I will tow you to the nearest port or anchorage (or place indicated) . . . .   **KP 1**

    I must get shelter or anchorage as soon as possible . . . .   **KP 2**

RG   You should send a boat to where I am to anchor or moor.

    RG 1   At what time shall I come into anchorage?

    You should proceed to anchorage in position indicated (lat… long…) . . . .   **RW**

    You should not proceed out of harbor/anchorage. . . . .   **RZ 1**

RH   There is no good holding ground in my area (or around lat… long…).

RI   There is good holding ground in my area (or around lat… long…).

    RI 1   Is there good holding ground in your area (or around lat… long…)?

Code		Meaning	Cross Reference

ENGINES—PROPELLER

Engines

Code		Meaning	Cross Reference
RJ		You should keep your engines ready.	
	RJ 1	You should have your engines ready as quickly as possible.	
	RJ 2	You should report when your engines are ready.	
	RJ 3	You should leave when your engines are ready.	
	RJ 4	At what time will your engines be ready?	
RK		My engines will be ready at time indicated.	
	RK 1	My engines are ready.	
RL		You should stop your engines immediately.	
	RL 1	You should stop your engines.	
RM		My engines are stopped.	
	RM 1	I am stopping my engines.	
	RM 2	I am obliged to stop my engines.	
RN		My engines are out of action.	
		I can only proceed with one engine	**IL 1**

Propeller

Code		Meaning	Cross Reference
RO		Propeller shaft is broken.	
	RO 1	My propeller is fouled by hawser or rope.	
	RO 2	I have lost my propeller.	
		I require immediate assistance; propeller shaft is broken	**CB 8**

LANDING—BOARDING

Landing

Code		Meaning	Cross Reference
***RP**		Landing here is highly dangerous.	
	***RP 1**	Landing here is highly dangerous. A more favorable location for landing is at position indicated.	
****RQ**		Interrogative or "the significance of the previous group should be read as a question".	
***RR**		This is the best place to land.	

Code		Meaning	Cross Reference

Landing

*RR 1   Lights will be shown or flag waved at the best landing place.

Boat should endeavor to land where flag is waved or light is shown . . . .   **DC**

Boats are not allowed to land (after time indicated). . . . .   **DD 1**

*   Reference is made to landing signals prescribed by the International Convention for the Safety of Life at Sea, 1974 (Regulation 16(b), Chapter V), for the guidance of small boats with crews or persons in distress.

**   Procedural signal.

Boarding

RS   No one is allowed on board.

You should stop or heave to, I am going to board you . . . .   **SQ 3**

MANEUVERS

RT   Stop carrying out your intentions and watch for my signals . . . .   **X**

    RT 1   What maneuvers do you intend to carry out?

RU   Keep clear of me; I am maneuvering with difficulty . . . .   **D**

    RU 1   I am carrying out maneuvering trials.

PROCEED—UNDERWAY

Proceed

RV   You should proceed (to place indicated if necessary).

    RV 1   You should proceed to destination.

    RV 2   You should proceed into port.

    RV 3   You should proceed to sea.

RW   You should proceed to anchorage in position indicated (lat… long…).

RX   You should proceed at time indicated.

RY   You should proceed at slow speed when passing me (or vessels making this signal).

You should proceed to the rescue of vessel (or ditched aircraft) in lat… long . . . .   **GP**

You should proceed to lat… long… to pick up survivors . . . .   **HN**

You should proceed with great caution . . . .   **NE**

SECTION 4.—MANEUVERS

Code	Meaning	Cross Reference

Proceed

	You should proceed with great caution; the coast is dangerous.	**NE 1**
	You should proceed with great caution; submarines are exercising in this area	**NE 2**
	You should proceed with great caution; there is a boom across	**NE 3**
	You should proceed with great caution; keep clear of firing range	**NE 4**
	You should proceed with great caution; hostile vessel sighted (in lat... long...)	**NE 5**
	You should proceed with great caution; hostile submarine sighted (in lat... long...).	**NE 6**
	You should proceed with great caution; hostile aircraft sighted (in lat... long...)	**NE 7**
RZ	You should not proceed (to place indicated if necessary).	
RZ 1	You should not proceed out of harbor/anchorage.	
	All vessels should proceed to sea as soon as possible owing to danger in port.	**UL**
SA	I can proceed at time indicated.	
SB	I am proceeding to the position of accident.	
	I am (or vessel indicated is) proceeding to your assistance.	**CP**
	I am proceeding to the assistance of vessel/aircraft in distress (lat... long...)	**CR**
	I am proceeding to the position of accident at full speed. Expect to arrive at time indicated . .	**FE**
	Are you proceeding to the position of accident? If so, when do you expect to arrive?.	**FE 1**
	I cannot proceed to the rescue owing to weather. You should do all you can	**GQ**
	I will try to proceed by my own means but I request you to keep in contact with me by... (Complements Table 1, Chapter 2, Section 10, Page 104)	**IJ**
	I can proceed at... (number) knots.	**IK**
	I can only proceed at slow speed	**IL**
	I can only proceed with one engine	**IL 1**
	I am unable to proceed under my own power	**IL 2**
	Are you in a condition to proceed	**IL 3**
	I have placed the collision mat. I can proceed without assistance.	**KA 1**

Underway

SC	I am underway.	
SC 1	I am ready to get underway.	
SC 2	I shall get underway as soon as the weather permits.	

Code		Meaning	Cross Reference

Underway

SD		I am not ready to get underway.	
SF		Are you (or vessel indicated) underway?	
	SF 1	Are you ready to get underway?	
	SF 2	At what time will you be underway?	

SPEED

SG		My present speed is… (number) knots.	
SJ		My maximum speed is… (number) knots.	
SL		What is your present speed?	
	SL 1	What is your maximum speed?	
		The speed of my aircraft in relation to the surface of the earth is… (knots or kilometers per hour)..	**BQ**
		What is the speed of your aircraft in relation to the surface of the earth?.................	**BQ 1**
		My present position, course, and speed are lat… long… ,… , knots....................	**EV**
		What are your present position, course, and speed?................................	**EV 1**
		Will vessels in my immediate vicinity (or in the vicinity of lat… long…) please indicate position, course, and speed..	**FB**
		I can only proceed at slow speed..	**IL**
		Maximum speed in tow is… (number) knots......................................	**LH**
		I am increasing speed...	**LI**
		Increase speed...	**LI 1**
		I am reducing speed..	**LJ**
		Reduce speed..	**LJ 1**
		You should proceed at slow speed when passing me (or vessels making this signal)	**RY**
		Take the way off your vessel	**SP**
		My vessel is stopped and making no way through the water........................	**SP 1**
SM		I am undergoing speed trials.	

Code		*Meaning*	*Cross Reference*

STOP—HEAVE TO

SN		You should stop immediately. Do not scuttle. Do not lower boats. Do not use the wireless. If you disobey I shall open fire on you.	
SO		You should stop your vessel instantly	**L**
	SO 1	You should stop. Head offshore.	
	SO 2	You should remain where you are.	
SP		Take the way off your vessel.	
	SP 1	My vessel is stopped and making no way through the water	**M**
SQ		You should stop or heave to.	
	SQ 1	You should stop or heave to, otherwise I shall open fire on you.	
	SQ 2	You should stop or heave to; I am going to send a boat.	
	SQ 3	You should stop or heave to; I am going to board you.	
		You should heave to or anchor until pilot arrives.	**UB**
		I am (or vessel indicated is) stopped in thick fog	**XP**

CHAPTER 2

SECTION 5: MISCELLANEOUS

CARGO—BALLAST

ST		What is your cargo?	
SU		My cargo is agricultural products.	
	SU 1	My cargo is coal.	
	SU 2	My cargo is dairy products.	
	SU 3	My cargo is fruit products.	
	SU 4	My cargo is heavy equipment/machinery.	
	SU 5	My cargo is livestock.	
	SU 6	My cargo is lumber.	
	SU 7	My cargo is oil/petroleum products.	
	SU 8	I have a general cargo.	
SV		I am not seaworthy due to shifting of cargo or ballast.	
SW		I am taking in, or discharging, or carrying dangerous goods	**B**
SX		You should not discharge oil or oily mixture.	
SY		The discharge of oil or oily mixture in this area is prohibited within… (number) miles from the nearest land.	

CREW—PERSONS ON BOARD

SZ	Total number of persons on board is…	
TA	I have left… (number) men on board.	
TB	… (number) persons have died.	
TC	… (number) persons are sick.	
	I am alighting (in position indicated if necessary) to pick up crew of vessel/aircraft	**AV**
	I cannot alight but I can lift crew	**AZ**
	You cannot alight on the deck. Can you lift crew?	**BA 1**
	Boats(s)/raft(s) should approach vessel as near as possible to take off persons.	**DA**
	All persons saved	**GZ**
	All persons lost	**GZ 1**
	I (or rescue vessel/aircraft) have rescued… (number) injured persons.	**HA**
	Can I transfer rescued persons to you?	**HD**

Code		Meaning	Cross Reference

FISHERY‡

TD		I am a fish catch carrier boat.	
	TD 1	I am a mother ship for fishing vessel(s).	
	TD 2	Are you a fishing vessel?	
TE		I am bottom trawling.	
	TE 1	I am trawling with a floating trawl.	
	TE 2	I am long-line fishing.	
	TE 3	I am fishing with towing lines.	
	TE 4	I am engaged in two-boat fishing operation.	
	TE 5	I am drifting on my nets.	
	TE 6	In what type of fishing are you engaged?	
TF		I am shooting purse seine.	
	TF 1	I am shooting drift nets.	
	TF 2	I am shooting seine net.	
	TF 3	I am shooting trawl.	
	TF 4	I am shooting long lines.	
TG		I am hauling purse seine.	
	TG 1	I am hauling drift nets.	
	TG 2	I am hauling seine net.	
	TG 3	I am hauling trawl.	
	TG 4	I am hauling long lines.	
TH		You should navigate with caution. Small fishing boats are within… (number) miles of me.	
TI		You should navigate with caution. You are drifting towards my set of nets.	
TJ		You should navigate with caution. There are nets with a buoy in this area.	
		There is fishing gear in the direction you are heading (or in direction indicated—Complements Table 3, Chapter 2, Section 10, Page 104)	NB
TK		Is there fishing gear set up on my course?	
TL		My gear is close to the surface in a direction… (Complements Table 3, Chapter 2, Section 10, Page 104) for a distance of… miles.	

---

‡ Displaying any of the signals in this section does not relieve vessels from compliance with the International Regulations for Preventing Collisions at Sea 1972.

Code		*Meaning*	*Cross Reference*
TM		My gear is well below the surface in a direction… (Complements Table 3, Chapter 2, Section 10, Page 104) for a distance of… miles.	
TN		In what direction, distance, and depth does your fishing gear extend?	
TO		I have a mine in my sweep (or net).	
TP		Fishing gear has fouled my propeller.	
TQ		You have caught my fishing gear.	
	TQ 1	It is necessary to haul in fishing gear for disentangling.	
	TQ 2	I am clearing the fishing gear.	
	TQ 3	You should take measures to recover the fishing gear.	
TS		You should take the following action with your warps:	
	TS 1	Veer the port (stern) warp.	
	TS 2	Veer the starboard (fore) warp.	
	TS 3	Veer both warps.	
	TS 4	Stop veering.	
	TS 5	Haul the port (stern) warp.	
	TS 6	Haul the starboard (fore) warp.	
	TS 7	Haul both warps.	
	TS 8	Stop hauling.	
	TS 9	You may haul your warps; the trawl is clear.	
TU		I have to cut the warps. The trawls are entangled.	
	TU 1	Give me your warp. I shall transfer your fishing gear on it.	
	TU 2	Your warps are under mine.	
	TU 3	Both my warps have parted.	
	TU 4	My starboard (fore) warp has parted.	
	TU 5	My port (stern) warp has parted.	
TV		Fishing in this area is prohibited.	
	TV 1	Trawling in this area is dangerous because there is an obstruction.	
TW		Attention. You are in the vicinity of prohibited fishery limits.	
TX		A fishery protection (or fishery assistance) vessel is in lat… long…	
TY		I request assistance from fishery protection (or fishery assistance) vessel.	
TZ		Can you offer assistance? (Complements Table 2, Chapter 2, Section 10, Page 104).	

Code		Meaning	Cross Reference

PILOT

(See also Navigating and Steering instructions in Chapter 2, Section 3, Page 73.)

UA		Pilot will arrive at time indicated.	
UB		You should heave to or anchor until pilot arrives.	
		I have a pilot on board	**H**
UC		Is a pilot available in this place (or place indicated)?	
		I require a pilot.	**G**
UE		Where can I get a pilot (for destination indicated if necessary)?	
UF		You should follow pilot boat (or vessel indicated).	
UG		You should steer in my wake.	
		You should follow in my wake (or wake of vessel indicated)	**PM**
		You should go ahead and lead the course	**PM 1**
UH		Can you lead me into port?	
UI		Sea is too rough; pilot boat cannot get off to you.	
UJ		Make a starboard lee for the pilot boat.	
	UJ 1	Make a port lee for the pilot boat.	
UK		Pilot boat is most likely on bearing… from you.	
	UK 1	Have you seen the pilot boat?	
		Is radar pilotage being effected in this port (or port indicated)?	**OL**

PORT—HARBOR

UL		All vessels should proceed to sea as soon as possible owing to danger in port.	
UM		The harbor (or port indicated) is closed to traffic.	
		You should not proceed out of harbor/anchorage	**RZ 1**
UN		You may enter the harbor immediately (or at time indicated).	
	UN 1	May I enter harbor?	
	UN 2	May I leave harbor?	
UO		You must not enter harbor.	
UP		Permission to enter harbor is urgently requested. I have an emergency case.	

Code	Meaning	Cross Reference
	You should proceed into port	**RV 2**
	Can you lead me into port?	**UH**
UQ	You should wait outside the harbor (or river mouth).	
UQ 1	You should wait outside the harbor until daylight.	
UR	My estimated time of arrival (at place indicated) is (time indicated).	
UR 1	What is your estimated time of arrival (at place indicated)?	

MISCELLANEOUS

Code	Meaning	
US	Nothing can be done until time indicated.	
US 1	Nothing can be done until daylight.	
US 2	Nothing can be done until tide has risen.	
US 3	Nothing can be done until visibility improves.	
US 4	Nothing can be done until weather moderates.	
US 5	Nothing can be done until draft is lightened.	
US 6	Nothing can be done until tugs have arrived.	
UT	Where are you bound for?	
UT 1	Where are you coming from?	
UU	I am bound for…	
UV	I am coming from…	
*UV 1	I am conducting innocent passage in the territorial sea.	
*UV 2	Your course leads into an area of the territorial sea in which the right of innocent passage is temporarily suspended.	
*UV 3	You should leave the area of the territorial sea in which the right of innocent passage is temporarily suspended.	
*UV 4	You are violating the conditions of innocent passage through the territorial sea (as indicated in the table of complements below). Request you comply with the conditions of innocent passage through the territorial sea.	
*UV 5	I am not violating the conditions of innocent passage through the territorial sea (as indicated in the table of complements below).	
*UV 6	I have ceased violating the conditions of innocent passage through the territorial sea (as indicated in the table of complements below).	
*UV 7	Having disregarded our request for compliance with the conditions of innocent passage through the territorial sea, you are required to leave the territorial sea immediately.	

Code	Meaning	Cross Reference

*UV 8   I am conducting transit passage through an international Strait.

*UV 9   I am exercising freedom of navigation.

*   Signals **UV 1 - UV 9** are not mandatory. Use of these signals is not a precondition to the exercise of the right of innocent passage or freedom of navigation.

Conditions of innocent passage… (Complements 0-9 corresponding to the following table):

0   By threatening or using force against our/your sovereignty, territorial integrity, or political independence.

1   By exercising or practicing with weapons.

2   By engaging in acts aimed at collecting information to the prejudice of our/your defense or security.

3   By engaging in acts of propaganda aimed at affecting our/your defense or security.

4   By engaging in the launching, landing or taking on board of aircraft or a military device.

5   By engaging in the loading or unloading of a commodity, currency or person contrary to the customs, fiscal, immigration or sanitary laws or regulations of our/your country.

6   By engaging in willful and serious pollution.

7   By engaging in fishing activities.

8   By engaging in research or survey activities.

9   By engaging in acts aimed at interfering with our/your systems of communication or other facilities or installations.

UW   I wish you a pleasant voyage.

UW 1   Thank you very much for your cooperation. I wish you a pleasant voyage.

UW 2   Welcome!

UW 3   Welcome Home!

UX   No information available.

I am unable to answer your question . . . . **YK**

Exercises

UY   I am carrying out exercises. Please keep clear of me.

Bunkers

UZ   I have bunkers for… (number) hours.

VB   Have you sufficient bunkers to reach port?

VC   Where is the nearest place at which fuel oil is available?

VC 1   Where is the nearest place at which diesel oil is available?

Code	Meaning	Cross Reference

Bunkers

Code	Meaning	Cross Reference
VC 2	Where is the nearest place at which coal is available?	
VD	Bunkers are available at place indicated (or lat… long…).	

Fumigation

| **VE** | I am fumigating my vessel. | |
| | No one is allowed on board . . . . | **RS** |

Identification

| | What is the name or identity signal of your vessel (or station)? . . . . | **CS** |
| **VF** | You should hoist your identity signal. | |

CHAPTER 2

SECTION 6: METEOROLOGY—WEATHER

CLOUDS

VG   The coverage of low clouds is… (number of octants or eighths of sky covered).

VH   The estimated height of base of low clouds in hundreds of meters is…

VI   What is the coverage of low clouds in octants (eighths of sky covered)?

         VI 1   What is the estimated height of base of low clouds in hundreds of meters?

GALE—STORM—TROPICAL STORM

Gale

VJ   Gale (wind force Beaufort 8-9) is expected from direction indicated (Complements Table 3, Chapter 2, Section 10, Page 104).

Storm

VK   Storm (wind force Beaufort 10 or above) is expected from direction indicated (Complements Table 3, Chapter 2, Section 10, Page 104).

Tropical Storm

VL   Tropical storm (cyclone, hurricane, typhoon) is approaching. You should take appropriate precautions.

VM   Tropical storm is centered at… (time indicated) in lat… long… on course… , speed… knots.

VN   Have you latest information of the tropical storm (near lat… long… if necessary)?

Very deep depression is approaching from direction indicated (Complements Table 3, Chapter 2, Section 10, Page 104). . . . . **WT**

There are indications of an intense depression forming in lat… long. **WU**

ICE—ICEBERGS

Ice

VO   Have you encountered ice?

VP   What is the character of ice, its development, and the effects on navigation?

VQ   Character of ice:

         VQ 0   No ice.

         VQ 1   New ice (ice crystals, slush or sludge, pancake ice or ice rind).

         VQ 2   Young fast ice (5-15 cms thick or rotten fast ice).

Code	Meaning	Cross Reference

Ice

VQ 3   Open drift ice (not more than 5/8 of the water surface is covered by ice floes).

VQ 4   A compressed accumulation of sludge (a compressed mass of sludge or pancake ice, the ice cannot spread).

VQ 5   Winter fast ice (more than 15 cms in thickness).

VQ 6   Close drift ice (the area is covered by ice floes to a greater extent than 5/8).

VQ 7   Very close drift ice on open sea.

VQ 8   Pressure ice or big, vast, heavy ice floes.

VQ 9   Shore lead along the coast.

No information available .... **UX**

VR   Ice development:

VR 0   No change.

VR 1   Ice situation has improved.

VR 2   Ice situation has deteriorated.

VR 3   Ice has been broken up.

VR 4   Ice has opened or drifted away.

VR 5   New ice has been formed and/or the thickness of the ice has increased.

VR 6   Ice has been frozen together.

VR 7   Ice has drifted into the area or has been squeezed together.

VR 8   Warning of pressure ridges.

VR 9   Warning of hummocking or ice screwing.

No information available .... **UX**

VS   Effects of the ice on navigation:

VS 0   Unobstructed.

VS 1   Unobstructed for power-driven vessels built of iron or steel, dangerous for wooden vessels without ice protection.

VS 2   Difficult for low-powered vessels without the assistance of an icebreaker, dangerous for vessels of weak construction.

VS 3   Possible without icebreaker only for high-powered vessels of strong construction.

VS 4   Icebreaker assistance available in case of need.

VS 5   Proceed in channel without the assistance of icebreaker.

Code	Meaning	Cross Reference

Ice

	Code	Meaning	Cross Reference
	VS 6	Possible only with the assistance of an icebreaker.	
	VS 7	Icebreaker can give assistance only to ships strengthened for navigation in ice.	
	VS 8	Temporarily closed for navigation.	
	VS 9	Navigation has ceased.	
		No information available	**UX**
VT		Danger of ice accretion on superstructure (for example, black frost).	
	VT 1	I am experiencing heavy icing on superstructure.	
VU		I have seen icefield in lat… long…	
VV		Ice patrol ship is not on station.	
	VV 1	Ice patrol ship is on station.	

Icebergs

Code	Meaning
VW	I have seen icebergs in lat… long…
VX	I have encountered one or more icebergs or growlers (with or without position and time).
VY	One or more icebergs or growlers have been reported (with or without position and time).
VZ	Navigation is dangerous in the area around lat… long… owing to iceberg(s).
VZ 1	Navigation is dangerous in the area around lat… long… owing to floating ice.
VZ 2	Navigation is dangerous in the area around lat… long… owing to pack ice.

ICEBREAKER‡

Code	Meaning	Cross Reference
***WA**	Repeat word or group after…	
***WB**	Repeat word or group before…	
WC	I am (or vessel indicated is) fast on ice and require(s) icebreaker assistance.	
WC 1	Icebreaker is being sent to your assistance.	
	I require assistance in the nature of icebreaker	**CD 9**
WD	Icebreaker is not available.	
WD 1	Icebreaker cannot render assistance at present.	

―――――――――――――――――――――――

‡ Special single letter signals for use between icebreakers and assisted vessels can be found in Chapter 1, Section 10, Pages 24 and 25.

Code	*Meaning*	*Cross Reference*
WE	Navigation channel is being kept open by icebreaker.	
WF	I can give icebreaker support only up to lat... long...	
WG	Open channel or open area is in the direction in which aircraft is flying.	
WH	I can only assist if you will make all efforts to follow.	
WI	At what time will you follow at full speed?	
WJ	The convoy will start at time indicated from here (or from lat... long...).	
WK	You (or vessel indicated) will be number... in convoy.	
WL	Icebreaker is stopping work during darkness.	
*WM	Icebreaker support is now commencing. Use special icebreaker support signals and keep continuous watch for sound, visual, or radiotelephony signals.	
WN	Icebreaker is stopping work for... (number) hours or until more favorable conditions arise.	
WO	Icebreaker support is finished. Proceed to your destination.	
	You should go astern..	QL

* Procedural signals.

ATMOSPHERIC PRESSURE—TEMPERATURE

Atmospheric Pressure

WP		Barometer is steady.
	WP 1	Barometer is falling rapidly.
	WP 2	Barometer is rising rapidly.
WQ		The barometer has fallen... (number) millibars during the past three hours.
WR		The barometer has risen... (number) millibars during the past three hours.
WS		Corrected atmospheric pressure at sea level is... (number) millibars.
	WS 1	State corrected atmospheric pressure at sea level in millibars.
WT		Very deep depression is approaching from direction indicated (Complements Table 3, Chapter 2, Section 10, Page 104).
WU		There are indications of an intense depression forming in lat... long...

Temperature

WV	The air temperature is sub-zero (centigrade).

Code	Meaning	Cross Reference

Temperature

WV 1   The air temperature is expected to be sub-zero (centigrade).

SEA—SWELL

Sea

WW   What are the sea conditions in your area (or around lat... long...)?

WX   The true direction of the sea in tens of degrees is... (number following indicates tens of degrees).

WY   The state of the sea is... (Complements 0-9 corresponding to the following table which measures wave height):

0 Calm (glassy)	0m	0 ft
1 Calm (rippled)	0-0.1m	0-$^1/_3$ ft
2 Smooth (wavelets)	0.1-0.5m	$^1/_3$-1 $^2/_3$ ft
3 Slight	0.5-1.2 m	1 $^2/_3$-4 ft
4 Moderate	1.25-2.5m	4-8 ft
5 Rough	2.5-4m	8-13 ft
6 Very rough	4-6m	13-20 ft
7 High	6-9m	20-30 ft
8 Very high	9-14m	30-45 ft
9 Phenomenal	over 14m	over 45 ft

WZ   What are the forecast sea conditions in my area (or area around lat... long...)?

XA   The true direction of the sea in tens of degrees is expected to be... (number following indicates tens of degrees).

XB   The state of the sea is expected to be... (Complements 0-9 as in the table above).

Swell

XC   What are the swell conditions in your area (or area around lat... long...)?

XD   The true direction of the swell in tens of degrees is... (number following indicates tens of degrees).

XE   The state of the swell is... (Complements 0-9 corresponding to the following table):

Code	*Meaning*	*Cross Reference*

0 No swell....................
1 short or middle......... } weak—approximate height <2 m (6 ft.)
2 Long.........................

3 Short........................
4 Middle.................... } moderate—approximate height 2-4 m (6-12 ft.)
5 Long.......................

6 Short......................
7 Middle.................... } high—approximate height > 4 m (12 ft.)
8 Long.......................
9 Confused

XF What are the forecast swell conditions in my area (or area around lat… long…)?

XG The true direction of the swell in tens of degrees is expected to be… (number following indicates tens of degrees).

XH The state of the swell is expected to be… (Complements 0-9 as in the table above).

Tsunami (phenomenal wave) is expected. You should take appropriate precautions. . . . .   **ND**

VISIBILITY—FOG

XI Indicate visibility.

XJ Visibility is… (number) tenths of nautical miles.

XK Visibility is variable between… and… (maximum and minimum in tenths of nautical miles).

XL Visibility is decreasing.

    XL 1 Visibility is increasing.

    XL 2 Visibility is variable.

XM What is the forecast visibility in my area (or area around lat… long…)?

XN Visibility is expected to be… (number) tenths of nautical miles.

XO Visibility is expected to decrease.

    XO 1 Visibility is expected to increase.

    XO 2 Visibility is expected to be variable.

XP I am (or vessel indicated is) stopped in thick fog.

    XP 1 I am entering zone of restricted visibility.

Code	Meaning	Cross Reference

WEATHER—WEATHER FORECAST

XQ   What weather are you experiencing?

XR   Weather is good.

    XR 1   Weather is bad.

    XR 2   Weather is moderating.

    XR 3   Weather is deteriorating.

XS   Weather report is not available.

XT   Weather expected is bad.

    XT 1   Weather expected is good.

    XT 2   No change is expected in the weather.

    XT 3   What weather is expected?

XU   You should wait until the weather moderates.

    XU 1   I will wait until the weather moderates.

XV   Please give weather forecast for my area (or area around lat… long…) in *MAFOR Code.

  * MAFOR is the prefix used to identify an International Coded Weather Forecast for Shipping.

WIND

XW   What is the true direction and force of wind in your area (or area around lat… long…)?

XX   True direction of wind is… (Complements Table 3, Chapter 2, Section 10, Page 104).

XY   Wind force is Beaufort Scale… (numerals 0-12).

XZ   What is the wind doing?

    XZ 1   The wind is backing.

    XZ 2   The wind is veering.

    XZ 3   The wind is increasing.

    XZ 4   The wind is squally.

    XZ 5   The wind is steady in force.

    XZ 6   The wind is moderating.

YA   What wind direction and force is expected in my area (or area around lat… long…)?

Code	Meaning	Cross Reference
YB	True direction of wind is expected to be… (Complements Table 3, Chapter 2, Section 10, Page 104).	
YC	Wind force expected is Beaufort Scale… (numerals 0-12).	
YD	What is the wind expected to do?	
YD 1	The wind is expected to back.	
YD 2	The wind is expected to veer.	
YD 3	The wind is expected to increase.	
YD 4	The wind is expected to become squally.	
YD 5	The wind is expected to remain steady in force.	
YD 6	The wind is expected to moderate.	

CHAPTER 2

SECTION 7: ROUTING OF SHIPS

Code	Meaning	Cross Reference
YG	You appear not to be complying with the traffic separation scheme.	
	It is dangerous to proceed on present course	**MY 2**
	You should proceed with great caution	**NE**
	You are in a dangerous position	**NG**

CHAPTER 2

SECTION 8: COMMUNICATIONS

Code	Meaning	Cross Reference

ACKNOWLEDGE—ANSWER

YH I have received the following from... (name or identity signal of vessel or station).

YI I have received the safety signal sent by... (name or identity signal).

YJ Have you received the safety signal sent by... (name or identity signal)?

YK I am unable to answer your question.

Received, or I have received your last signal . . . . **R***

*   Procedural signal.

CALLING

YL I will call you again at... hours (on... kHz or MHz).

YM Who is calling me?

CANCEL

YN Cancel my last signal/message.

My last signal was incorrect. I will repeat it correctly . . . . **ZP**

COMMUNICATE

I wish to communicate with you by... (Complements Table 1, Chapter 2, Section 10, Page 104). **K***

I wish to communicate with you . . . . **K**

YO I am going to communicate by... (Complements Table 1, Chapter 2, Section 10, Page 104).

YP I wish to communicate with vessel or coast station (identity signal) by... (Complements Table 1, Chapter 2, Section 10, Page 104).

YQ I wish to communicate by... (Complements Table 1, Chapter 2, Section 10, Page 104) with vessel bearing... from me.

YR Can you communicate by... (Complements Table 1, Chapter 2, Section 10, Page 104)?

YS I am unable to communicate... (Complements Table 1, Chapter 2, Section 10, Page 104).

YT I cannot read your... (Complements Table 1, Chapter 2, Section 10, Page 104).

**YU I am going to communicate with your station by means of the International Code of Signals.

**YV The groups which follow are from the International Code of Signals.

    YV 1 The groups which follow are from the local code.

SECTION 8.—COMMUNICATIONS

Code	Meaning	Cross Reference
YW	I wish to communicate by radiotelegraphy on frequency indicated.	
YX	I wish to communicate by radiotelephony on frequency indicated.	
YY	I wish to communicate by VHF radiotelephony on channel indicated.	
YZ	The words which follow are in plain language.	
ZA	I wish to communicate with you in… (language indicated by following complements).	
ZA 0	Dutch	
ZA 1	English	
ZA 2	French	
ZA 3	German	
ZA 4	Greek	
ZA 5	Italian	
ZA 6	Japanese	
ZA 7	Norwegian	
ZA 8	Russian	
ZA 9	Spanish	
ZB	I can communicate with you in language indicated (complements as above).	
ZC	Can you communicate with me in language indicated (complements as above)?	
ZD	Please communicate the following to all shipping in the vicinity.	
ZD 1	Please report me to Coast Guard New York.	
ZD 2	Please report me to Lloyd's London.	
ZD 3	Please report me to Minmorflot Moscow.	
ZD 4	Please report me to MSA Tokyo.	
ZE	You should come within visual signal distance.	
	You should keep within visual signal distance from me (or vessel indicated).	**PR 2**
	I have established communications with the aircraft in distress on 2182 kHz	**BC**
	Can you communicate with the aircraft?	**BC 1**
	I have established communications with the aircraft in distress on… kHz	**BD**
	I have established communications with the aircraft in distress on… MHz	**BE**

* With one numeral.
** The abbreviation INTERCO may also be used to mean: "International Code group(s) follow(s)".

Code		*Meaning*	*Cross Reference*

EXERCISE

ZF		I wish to exercise signals with you by… (Complements Table 1, Chapter 2, Section 10, Page 104).	
ZG		It is not convenient to exercise signals.	
ZH		Exercise had been completed.	

RECEPTION—TRANSMISSION

ZI		I can receive but not transmit by… (Complements Table 1, Chapter 2, Section 10, Page 104).	
ZJ		I can transmit but not receive by… (Complements Table 1, Chapter 2, Section 10, Page 104).	
ZK		I cannot distinguish your signal. Please repeat it by… (Complements Table 1, Chapter 2, Section 10, Page 104).	
ZL		Your signal has been received but not understood.	
		I cannot read your… (Complements Table 1, Chapter 2, Section 10, Page 104).	YT
ZM		You should send (or speak) more slowly.	
	ZM 1	Shall I send (or speak) more slowly?	
ZN		You should send each word or group twice.	
ZO		You should stop sending.	
	ZO 1	Shall I stop sending?	

REPEAT

ZP		My last signal was incorrect. I will repeat it correctly.	
ZQ		Your signal appears incorrectly coded. You should check and repeat the whole.	
ZR		Repeat the signal now being made to me by vessel (or coast station)… (name or identity signal).	

CHAPTER 2

SECTION 9: INTERNATIONAL HEALTH REGULATIONS

Code		Meaning	Cross Reference

PRATIQUE MESSAGES

Code		Meaning	Cross Reference
ZS		My vessel is "healthy" and I request free pratique	**Q**
		*I require health clearance	**QQ**
ZT		My Maritime Declaration of Health has negative answers to the six Health Questions.	
ZU		My Maritime Declaration of Health has a positive answer to Health Question(s)… (Health Questions are indicated by complements 1-6).	
ZV		I believe I have been in an infected area during the last thirty days.	
ZW		I require Port Medical Officer.	
	ZW 1	Port Medical Officer will be available at (time indicated).	
ZX		You should make the appropriate pratique signal.	
ZY		You have pratique.	
ZZ		You should proceed to anchorage for health clearance (at place indicated).	
	ZZ 1	Where is the anchorage for health clearance?	
		I have a doctor on board.	**AL**
		Have you a doctor?.	**AM**

* By night a red light over a white light may be shown, where it can best be seen, by vessels requiring health clearance. These lights should only be about 6 feet apart, should be exhibited within the precincts of a port and should be visible all around the horizon as nearly as possible.

CHAPTER 2

SECTION 10: TABLES OF COMPLEMENTS

Table 1

1. Morse signaling by hand flags or arms
2. Loud hailer (megaphone)
3. Morse signaling lamp
4. Sound signals

Table 2

0. Water
1. Provisions
2. Fuel
3. Pumping equipment
4. Firefighting appliances
5. Medical assistance
6. Towing
7. Survival craft
8. Vessel to stand by
9. Icebreaker

Table 3

0. Direction unknown (or calm)
1. Northeast
2. East
3. Southeast
4. South
5. Southwest
6. West
7. Northwest
8. North
9. All directions (or confused or variable)

CHAPTER 3
MEDICAL SIGNAL CODE

CHAPTER 3

SECTION 1: EXPLANATION AND INSTRUCTIONS

General

1. Medical advice should be sought and given in plain language whenever it is possible but, if language difficulties are encountered, this Code should be used.

2. Even when plain language is used, the text of the Code and the instructions should be followed as far as possible.

3. Reference is made to the procedure signals **"C"**, **"N"**, or **"NO"** and **"RQ"** which, when used after the main signal, change its meaning into affirmative, negative and interrogative, respectively. (See Chapter 1, Section 6, Paragraph 3.(j), Page 11.)

> *Example:*
>
> **"MFE N"** = "Bleeding is not severe".
> **"MFE RQ"** = "Is bleeding severe?"

INSTRUCTIONS TO MASTERS

Standard method of case description

1. The master should make a careful examination of the patient and should try to collect, as far as possible, information covering the following subjects:

(a) Description of the patient (Chapter 3, Section 2., Page 109);

(b) Previous health (Chapter 3, Section 2., Page 110);

(c) Localization of symptoms, diseases, or injuries (Chapter 3, Section 2., Page 110);

(d) General symptoms (Chapter 3, Section 2., Page 110);

(e) Particular symptoms (Chapter 3, Section 2., Page 114);

(f) *Diagnosis (Chapter 3, Section 3., Page 125).

2. Such information should be coded by choosing the appropriate groups from the corresponding sections of this chapter. It would help the recipients of the signal if the information is transmitted in the order stated in Paragraph 1.

3. Chapter 3, Section 2., Page 109, contains signals which can be used independently, i.e. with or without the description of the case.

4. After a reply from the doctor has been received and the instructions therein followed, the master can give a progress report by using signals from Chapter 3, Section 2., Page 123.

INSTRUCTIONS TO DOCTORS

1. Additional information can be requested by using Chapter 3, Section 3, Page 125.

> *Example:*
>
> **"MQB"** = "I cannot understand your signal, please use standard method of case description".

2. For diagnosis*, Chapter 3, Section 3., Page 125, should be used.

> *Example:*
>
> **"MQE 26"** = "My probable diagnosis is cystitis".

3. Prescribing should be limited to the "List of Medicaments" which comprises Table M-3 in Chapter 3, Section 4, Pages 134 and 135, of the Code.

4. For special treatment, signals from Chapter 3, Section 3., Pages 125 through 127, should be used.

> *Example:*
>
> **"MRP 4"** = "Apply ice-cold compress and renew every 4 hours".

5. When prescribing a medicament (Chapter 3, Section 3., page 127) three signals should be used as follows:

   (a) the first (Chapter 3, Section 3., Page 127, and Table M-3 in Chapter 3, Section 4, Pages 134 and 135) to signify the medicament itself.

---

* Chapter 3, Section 3, Page 125 , "Diagnosis", can be used by both the master ("request for medical assistance") and the doctor ("medical advice").

Example:

"MTD 32" = "You should give aspirin tablets".

(b) the second (Chapter 3, Section 3, Page 127) to signify the method of administration and dose.

Example:

"MTI 2" = "You should give by mouth 2 tablets/capsules".

(c) the third (Chapter 3, Section 3, Page 127) to signify the frequency of the dose.

Example:

"MTQ 8" = "You should repeat every 8 hours".

6. The frequency of external applications is coded in Chapter 3, Section 3, Page 128.

Example:

"MTU 4" = "You should apply every 4 hours".

7. Advice concerning diet can be given by using signals from Chapter 3, Section 3, Page 128.

Example:

"MUC" = "Give water only in small quantities".

EXAMPLES

As an example, two cases of request for assistance and the corresponding replies are drafted below:

CASE ONE

Request for medical assistance

"I have a male age (44) years. Patient has been ill for (2) days. Patient has suffered from (bronchitis acute). Onset was sudden. Patient is delirious. Patient has fits of shivering. Temperature taken in mouth is (40). Pulse rate per minute is (110). The rate of breathing per minute is (30). Patient is in pain (chest). Part of the body affected is right (chest). Pain is increased on breathing. Patient has severe cough. Patient has blood-stained sputum. Patient has been given (penicillin injection) without effect. Patient has received treatment by medicaments in last (18) hours. My probable diagnosis is (pneumonia)."

Medical advice

"Your diagnosis is probably right. You should continue giving (penicillin injection). You should repeat every (12) hours. Put patient to bed lying down at absolute rest. Keep patient warm. Give fluid diet, milk, fruit juice, tea, mineral water. Give water very freely. Refer back to me in (24) hours or before if patient worsens."

CASE TWO

Request for medical assistance

"I have a male aged (31) years. Patient has been ill for (3) hours. Patient has had no serious previous illness. Pulse rate per minute is (95). Pulse is weak. Patient is sweating. Patient is in pain in lumbar (kidney) region. The part affected is left lumbar (kidney) region. Pain is severe. Pain is increased by hand pressure. Bowels are regular."

Request for additional information

"I cannot make a diagnosis. Please answer the following question(s). Temperature taken in the mouth is (number). Pain radiates to groin and testicle. Patient has pain on passing water. Urinary functions normal. Vomiting is present."

Additional information

"Temperature taken in mouth is (37). Pain radiates to groin and testicle. Patient has pain on passing water. Patient is passing small quantities of urine frequently. Vomiting is absent. Patient has nausea."

Medical advice

"My probable diagnosis is kidney stone (renal colic). You should give morphine injection. You should give by subcutaneous injection (15) milligrams. Give water freely. Apply hot water bottle to lumbar (kidney) region. Patient should be seen by doctor when next in port."

CHAPTER 3

SECTION 2: REQUEST FOR MEDICAL ASSISTANCE

Code	Meaning	Cross Reference

REQUEST—GENERAL INFORMATION

Code	Meaning	Cross Reference
MAA	I request urgent medical advice.	
MAB	I request you to make rendezvous in position indicated.	
MAC	I request you to arrange hospital admission.	
MAD	I am . . . (indicate number) hours from the nearest port.	
MAE	I am converging on nearest port.	
MAF	I am moving away from nearest port.	
	I require medical assistance	**W**
	I have a doctor on board.	**AL**
	Have you a doctor?	**AM**
	I need a doctor	**AN**
	I need a doctor; I have severe burns	**AN 1**
	I need a doctor; I have radiation casualties	**AN 2**
	I require a helicopter urgently with a doctor	**BR 2**
	I require a helicopter urgently to pick up injured/sick person	**BR 3**
	Helicopter is coming to you now (or at time indicated) with a doctor	**BT 2**
	Helicopter is coming to you now (or at time indicated) to pick up injured/sick person	**BT 3**
	I have injured/sick person (or number of persons indicated) to be taken off urgently.	**AQ**
	You should send a helicopter/boat with a stretcher	**BS**
	A helicopter/boat is coming to take injured/sick	**BU**
	You should send injured/sick persons to me	**AT**

DESCRIPTION OF PATIENT

Code	Meaning	Cross Reference
MAJ	I have a male aged . . . (number) years.	
MAK	I have a female aged . . . (number) years.	
MAL	I have a female . . . (number) months pregnant.	
MAM	Patient has been ill for . . . (number) days.	
MAN	Patient has been ill for . . . (number) hours.	
MAO	General condition of the patient is good.	
MAP	General condition of the patient is serious.	

Code	Meaning	Cross Reference

MAQ   General condition of the patient is unchanged.

MAR   General condition of the patient has worsened.

MAS   Patient has been given . . . (Table M-3 in Chapter 3, Section 4, Pages 134 and 135) with effect.

MAT   Patient has been given . . . (Table M-3 in Chapter 3, Section 4, Pages 134 and 135) without effect.

MAU   Patient has received treatment by medicaments in last . . . (indicate number) hours.

PREVIOUS HEALTH

MBA   Patient has suffered from . . . (Table M-2 in Chapter 3, Section 4, Page 133).

MBB   Patient has had previous operation . . . (Table M-2 in Chapter 3, Section 4, Page 133).

MBC   Patient has had no serious previous illness.

MBD   Patient has had no relevant previous injury.

LOCALIZATION OF SYMPTOMS, DISEASES, OR INJURIES

MBE   The whole body is affected.

MBF   The part of the body affected is . . . (Table M-1 in Chapter 3, Section 4, Page 130).

*MBG   The part of the body affected is right . . . (Table M-1 in Chapter 3, Section 4, Page 130).

*MBH   The part of the body affected is left . . . (Table M-1 in Chapter 3, Section 4, Page 130).

  * To be used when right and left side of the body or limb need to be differentiated.

GENERAL SYMPTOMS

MBP   Onset was sudden.

MBQ   Onset was gradual.

Temperature

MBR   Temperature taken in mouth is . . . (number).

MBS   Temperature taken in rectum is . . . (number).

MBT   Temperature in the morning is . . . (number).

MBU   Temperature in the evening is . . . (number).

Code	Meaning	Cross Reference

Temperature

MBV	Temperature is rising.	
MBW	Temperature is falling.	

Pulse

MBX	The pulse rate per minute is . . . (number).	
MBY	The pulse rate is irregular.	
MBZ	The pulse rate is rising.	
MCA	The pulse rate is falling.	
MCB	The pulse is weak.	
MCC	The pulse is too weak to count.	
MCD	The pulse is too rapid to count.	

Breathing

MCE	The rate of breathing per minute is . . . (number) (in and out being counted as one breath).	
MCF	The breathing is weak.	
MCG	The breathing is wheezing.	
MCH	The breathing is regular.	
MCI	The breathing is irregular.	
MCJ	The breathing is strenuous (noisy).	

Sweating

MCL	Patient is sweating.	
MCM	Patient has fits of shivering (chills).	
MCN	Patient has night sweats.	
MCO	Patient's skin is hot and dry.	
MCP	Patient is cold and clammy.	

Mental State and Consciousness

MCR	Patient is conscious.	
MCT	Patient is semiconscious but can be roused.	

Code	Meaning	Cross Reference

Mental State and Consciousness

Code	Meaning
MCU	Patient is unconscious.
MCV	Patient found unconscious.
MCW	Patient appears to be in a state of shock.
MCX	Patient is delirious.
MCY	Patient has mental symptoms.
MCZ	Patient is paralyzed . . . (Table M-1 in Chapter 3, Section 4, Page 130).
MDC	Patient is restless.
MDD	Patient is unable to sleep.

Pain

Code	Meaning
MDF	Patient is in pain . . . (Table M-1 in Chapter 3, Section 4, Page 130).
MDG	Pain is a dull ache.
MDJ	Pain is slight.
MDL	Pain is severe.
MDM	Pain is intermittent.
MDN	Pain is continuous.
MDO	Pain is increased by hand pressure.
MDP	Pain radiates to . . . (Table M-1 in Chapter 3, Section 4, Page 130).
MDQ	Pain is increased on breathing.
MDR	Pain is increased by action of bowels.
MDT	Pain is increased on passing water.
MDU	Pain occurs after taking food.
MDV	Pain is relieved by taking food.
MDW	Pain has no relation to taking food.
MDX	Pain is relieved by heat.
MDY	Pain has ceased.

Cough

Code	Meaning
MED	Cough is present.
MEF	Cough is absent.

Code	*Meaning*	*Cross Reference*

Bowels

MEG	Bowels are regular.	
MEJ	Patient is constipated and bowels last opened . . . (indicate number of days).	
MEL	Patient has diarrhea . . . (indicate number of times daily).	

Vomiting

MEM	Vomiting is present.	
MEN	Vomiting is absent.	
MEO	Patient has nausea.	

Urine

MEP	Urinary functions normal.	
MEQ	Urinary functions abnormal.	

Bleeding

MER	Bleeding is present . . . (Table M-1 in Chapter 3, Section 4, Page 130).	
MET	Bleeding is absent.	

Rash

MEU	A rash is present . . . (Table M-1 in Chapter 3, Section 4, Page 130).	
MEV	A rash is absent.	

Swelling

MEW	Patient has a swelling . . . (Table M-1 in Chapter 3, Section 4, Page 130).	
MEX	Swelling is hard.	
MEY	Swelling is soft.	
MEZ	Swelling is hot and red.	
MFA	Swelling is painful on hand pressure.	
MFB	Swelling is discharging.	
MFC	Patient has an abscess . . . (Table M-1 in Chapter 3, Section 4, Page 130).	
MFD	Patient has a carbuncle . . . (Table M-1 in Chapter 3, Section 4, Page 130).	

Code	*Meaning*	*Cross Reference*

PARTICULAR SYMPTOMS

Accidents, Injuries, Fractures, Suicide, and Poisons

	Bleeding is present . . . (Table M-1 in Chapter 3, Section 4, Page 130).	**MER**
MFE	Bleeding is severe.	
MFF	Bleeding is slight.	
MFG	Bleeding has been stopped by pad(s) and bandaging.	
MFH	Bleeding has been stopped by tourniquet.	
MFI	Bleeding has stopped.	
MFJ	Bleeding cannot be stopped.	
MFK	Patient has a superficial wound . . . (Table M-1 in Chapter 3, Section 4, Page 130).	
MFL	Patient has a deep wound . . . (Table M-1 in Chapter 3, Section 4, Page 130).	
MFM	Patient has penetrating wound . . . (Table M-1 in Chapter 3, Section 4, Page 130).	
MFN	Patient has a clean-cut wound . . . (Table M-1 in Chapter 3, Section 4, Page 130).	
MFO	Patient has a wound with ragged edges . . . (Table M-1 in Chapter 3, Section 4, Page 130).	
MFP	Patient has a discharging wound . . . (Table M-1 in Chapter 3, Section 4, Page 130).	
MFQ	Patient has contusion (bruising) . . . (Table M-1 in Chapter 3, Section 4, Page 130).	
MFR	Wound is due to blow.	
MFS	Wound is due to crushing.	
MFT	Wound is due to explosion.	
MFU	Wound is due to fall.	
MFV	Wound is due to gunshot.	
MFW	Patient has a foreign body in wound.	
MFX	Patient is suffering from concussion.	
MFY	Patient cannot move the arm . . . (Table M-1 in Chapter 3, Section 4, Page 130).	
MFZ	Patient cannot move the leg . . . (Table M-1 in Chapter 3, Section 4, Page 130).	
MGA	Patient has dislocation . . . (Table M-1 in Chapter 3, Section 4, Page 130).	
MGB	Patient has simple fracture . . . (Table M-1 in Chapter 3, Section 4, Page 130).	
MGC	Patient has compound fracture . . . (Table M-1 in Chapter 3, Section 4, Page 130).	
MGD	Patient has comminuted fracture . . . (Table M-1 in Chapter 3, Section 4, Page 130).	
MGE	Patient has attempted suicide.	

SECTION 2.—REQUEST FOR MEDICAL ASSISTANCE

Code	Meaning	Cross Reference

Code | *Meaning* | *Cross Reference*

Accidents, Injuries, Fractures, Suicide, and Poisons

MGF   Patient has cut throat.

MGG   Patient has superficial burn . . . (Table M-1 in Chapter 3, Section 4, Page 130).

MGH   Patient has severe burn . . . (Table M-1 in Chapter 3, Section 4, Page 130).

MGI   Patient is suffering from noncorrosive poisoning (no staining and burning of mouth and lips).

MGJ   Patient has swallowed corrosive (staining and burning of mouth and lips).

MGK   Patient has swallowed unknown poison.

MGL   Patient has swallowed a foreign body.

MGM   Emetic has been given with good results.

MGN   Emetic has been given without good results.

MGO   No emetic has been given.

MGP   Patient has had corrosive thrown on him . . . (Table M-1 in Chapter 3, Section 4, Page 130).

MGQ   Patient has inhaled poisonous gases, vapors, dust.

MGR   Patient is suffering from animal bite . . . (Table M-1 in Chapter 3, Section 4, Page 130).

MGS   Patient is suffering from snake bite . . . (Table M-1 in Chapter 3, Section 4, Page 130).

MGT   Patient is suffering from gangrene . . . (Table M-1 in Chapter 3, Section 4, Page 130).

Diseases of Nose and Throat

MGU   Patient has nasal discharge.

MGV   Patient has foreign body in nose.

MHA   Lips are swollen.

MHB   Tongue is dry.

MHC   Tongue is coated.

MHD   Tongue is glazed and red.

MHF   Tongue is swollen.

MHG   Patient has ulcer on tongue.

MHJ   Patient has ulcer in mouth.

MHK   Gums are sore and bleeding.

MHL   Throat is sore and red.

MHM   Throat has pinpoint white spots on tonsils.

Code	Meaning	Cross Reference

Diseases of Nose and Throat

Code	Meaning	Cross Reference
MHN	Throat has gray white patches on tonsils.	
MHO	Throat hurts and is swollen on one side.	
MHP	Throat hurts and is swollen on both sides.	
MHQ	Swallowing is painful.	
MHR	Patient cannot swallow.	
MHT	Patient has hoarseness of voice.	
	Patient has swallowed a foreign body	**MGL**
MHV	Patient has severe toothache.	

Diseases of Respiratory System

Code	Meaning	Cross Reference
MHY	Patient has pain in chest on breathing . . . (Table M-1 in Chapter 3, Section 4, Page 130).	
	Breathing is wheezing	**MCG**
MHZ	Breathing is deep.	
MIA	Patient has severe shortness of breath.	
MIB	Patient has asthmatical attack.	
	Cough is absent	**MEF**
MIC	Patient has severe cough.	
MID	Cough is longstanding.	
MIF	Patient is coughing up blood.	
MIG	Patient has no sputum.	
MIJ	Patient has abundant sputum.	
MIK	Sputum is offensive.	
MIL	Patient has bloodstained sputum.	
MIM	Patient has blueness of face.	

Diseases of the Digestive System

Code	Meaning	Cross Reference
MIN	Patient has tarry stool.	
MIO	Patient has clay-colored stool.	
	Patient has diarrhea . . . (indicate number of times daily)	**MEL**
MIP	Patient has diarrhea with frequent stools like rice water.	

Code	Meaning	Cross Reference

Diseases of the Digestive System

Code	Meaning	Cross Reference
MIQ	Patient is passing blood with stools.	
MIR	Patient is passing mucus with stools.	
	Patient has nausea	**MEO**
MIT	Patient has persistent hiccough.	
MIU	Patient has cramp pains and vomiting.	
	Vomiting is present.	**MEM**
	Vomiting is absent	**MEN**
MIV	Vomiting has stopped.	
MIW	Vomiting is persistent.	
MIX	Vomit is streaked with blood.	
MIY	Patient vomiting much blood.	
MIZ	Vomit is dark (like coffee grounds).	
MJA	Patient vomits any food and liquid given.	
MJB	Amount of vomit is . . . (indicate in deciliters: 1 deciliter equals one-sixth of a pint).	
MJC	Frequency of vomiting is . . . (indicate number) daily.	
MJD	Patient has flatulence.	
MJE	Wind has not been passed per anus for . . . (indicate number of hours).	
MJF	Wind is being passed per anus.	
MJG	Abdomen is distended.	
MJH	Abdominal wall is soft (normal).	
MJI	Abdominal wall is hard and rigid.	
MJJ	Abdominal wall is tender . . . (Table M-1 in Chapter 3, Section 4, Page 130).	
	Patient is in pain . . . (Table M-1 in Chapter 3, Section 4, Page 130)	**MDF**
	Patient has a swelling . . . (Table M-1 in Chapter 3, Section 4, Page 130)	**MEW**
MJK	Hernia is present.	
MJM	Hernia cannot be replaced.	
MJN	Hernia is painful and tender.	
MJO	Patient has bleeding hemorrhoids.	
MJP	Hemorrhoids cannot be reduced (put back in place).	

Code	Meaning	Cross Reference

Diseases of the Genitourinary System

	Patient is in pain . . . (Table M-1 in Chapter 3, Section 4, Page 130)	**MDF**
MJS	Patient has pain on passing water.	
MJT	Patient has pain in penis at end of passing water.	
MJU	Patient has pain spreading from abdomen to penis, testicles, or thigh.	
MJV	Patient is unable to hold urine (incontinent).	
MJW	Patient is unable to pass urine.	
MJX	Patient is passing small quantities of urine frequently.	
MJY	Amount of urine passed in 24 hours . . . (indicate number in deciliters: 1 deciliter equals one-sixth of a pint).	
	Urinary functions normal	**MEP**
MKA	Urine contains albumen.	
MKB	Urine contains sugar.	
MKC	Urine contains blood.	
MKD	Urine is very dark brown.	
MKE	Urine is offensive and may contain pus.	
MKF	Penis is swollen.	
MKH	Foreskin will not go back to normal position.	
MKI	Patient has swelling of testicles.	
MKJ	Shall I pass a catheter?	
MKK	I have passed a catheter.	
MKL	I am unable to pass a catheter.	

Diseases of the Nervous System and Mental Diseases

MKP	Patient has headache . . . (Table M-1 in Chapter 3, Section 4, Page 130).	
MKQ	Headache is throbbing.	
MKR	Headache is very severe.	
MKS	Head cannot be moved forwards to touch chest.	
MKT	Patient cannot feel pinprick . . . (Table M-1 in Chapter 3, Section 4, Page 130).	
MKU	Patient is unable to speak properly.	

118

SECTION 2.—REQUEST FOR MEDICAL ASSISTANCE

Code	Meaning	Cross Reference

Diseases of the Nervous System and Mental Diseases

Code	Meaning	Cross Reference
MKV	Giddiness (vertigo) is present.	
	Patient is paralyzed . . . (Table M-1 in Chapter 3, Section 4, Page 130)	**MCZ**
	Patient is conscious	**MCR**
	Patient is semiconscious but can be roused	**MCT**
	Patient is unconscious	**MCU**
MKW	Pupils are equal in size.	
MKX	Pupils are unequal in size.	
MKY	Pupils do not contract in a bright light.	
MKZ	Patient has no control over his bowels.	
MLA	Patient has fits associated with rigidity of muscles and jerking of limbs—indicate number of fits per 24 hours.	
	Patient has mental symptoms	**MCY**
MLB	Patient has delusions.	
MLC	Patient is depressed.	
	Patient is delirious	**MCX**
MLD	Patient is uncontrollable.	
	Patient has attempted suicide	**MGE**
MLE	Patient has had much alcohol.	
MLF	Patient has delirium tremens.	
MLG	Patient has bedsores . . . (Table M-1 in Chapter 3, Section 4, Page 130).	

Diseases of the Heart and Circulatory System

Code	Meaning	Cross Reference
	Patient is in pain . . . (Table M-1 in Chapter 3, Section 4, Page 130)	**MDF**
MLH	Pain has been present for . . . (indicate number of minutes).	
MLI	Pain in chest is constricting in character.	
MLJ	Pain is behind the breastbone.	
	Pain radiates to . . . (Table M-1 in Chapter 3, Section 4, Page 130)	**MDP**
	Patient has blueness of face	**MIM**
MLK	Patient has pallor.	

Code	Meaning	Cross Reference

Diseases of the Heart and Circulatory System

	The rate of breathing per minute is . . . (number) (in and out being counted as one breath)	**MCE**
	The pulse is weak	**MCB**
	The pulse rate is irregular	**MBY**
	The pulse is too weak to count	**MCC**
	The pulse is too rapid to count	**MCD**
MLL	Breathing is difficult when lying down.	
MLM	Swelling of legs that pits on pressure.	
MLN	Patient has varicose ulcer.	

Infectious and Parasitic Diseases

MLR	Rash has been present for . . . (indicate number of hours).	
MLS	Rash first appeared on . . . (Table M-1 in Chapter 3, Section 4, Page 130).	
MLT	Rash is spreading to . . . (Table M-1 in Chapter 3, Section 4, Page 130).	
MLU	Rash is fading.	
MLV	Rash is itchy.	
MLW	Rash is not itchy.	
MLX	Rash looks like general redness.	
MLY	Rash looks like blotches.	
MLZ	Rash looks like small blisters containing clear fluid.	
MMA	Rash looks like larger blisters containing pus.	
MMB	Rash is weeping (oozing).	
MMC	Rash looks like weals.	
MMD	Rash consists of rose-colored spots that do not blench on pressure.	
MME	Skin is yellow.	
	Patient has an abscess . . . (Table M-1 in Chapter 3, Section 4, Page 130)	**MFC**
MMF	Patient has buboes . . . (Table M-1 in Chapter 3, Section 4, Page 130).	
MMJ	Patient has been isolated.	
MMK	Should patient be isolated?	
MML	I have had (indicate number) similar cases.	

SECTION 2.—REQUEST FOR MEDICAL ASSISTANCE

Code	Meaning	Cross Reference

Infectious and Parasitic Diseases

	Meaning	Cross Reference
	Patient has diarrhea with frequent stools like rice water	**MIP**
	Patient has never been successfully vaccinated against smallpox	**MUT**
	Patient was last vaccinated . . . (date indicated)	**MUU**
	Patient has vaccination marks	**MUV**

Venereal Diseases (See also Diseases of Genitourinary System.)

Code	Meaning	Cross Reference
MMP	Patient has discharge from penis.	
MMQ	Patient has previous history of gonorrhea.	
MMR	Patient has single hard sore on penis.	
MMS	Patient has multiple sores on penis.	
	Patient has buboes . . . (Table M-1 in Chapter 3, Section 4, Page 130)	**MMF**
MMT	Patient has swollen glands in the groin.	
MMU	End of penis is inflamed and swollen.	

Diseases of the Ear

Code	Meaning	Cross Reference
	Patient is in pain . . . (Table M-1 in Chapter 3, Section 4, Page 130)	**MDF**
MMW	Patient has boil in ear(s).	
MMX	Patient has discharge of blood from ear(s).	
MMY	Patient has discharge of clear fluid from ear(s).	
MMZ	Patient has discharge of pus from ear(s).	
MNA	Patient has hearing impaired.	
MNB	Patient has foreign body in ear.	
	Giddiness (vertigo) is present	**MKV**
MNC	Patient has constant noises in ear(s).	

Diseases of the Eye

Code	Meaning	Cross Reference
	Patient is in pain . . . (Table M-1 in Chapter 3, Section 4, Page 130)	**MDF**
MNG	Patient has inflammation of eye(s).	
MNH	Patient has discharge from eye(s).	
MNI	Patient has foreign body embedded in the pupil area of the eye.	

CHAPTER 3.—MEDICAL SIGNAL CODE

Code	*Meaning*	*Cross Reference*

Diseases of the Eye

MNJ	Eyelids are swollen.	
MNK	Patient cannot open eyes (raise eyelids).	
MNL	Patient has foreign body embedded in the white of the eye.	
MNM	Patient has double vision when looking at objects with both eyes open.	
MNN	Patient has sudden blindness in one eye.	
MNO	Patient has sudden blindness in both eyes.	
	Pupils are equal in size	MKW
	Pupils are unequal in size	MKX
	Pupils do not contract in a bright light	MKY
	Patient has a penetrating wound . . . (Table M-1 in Chapter 3, Section 4, Page 130)	MFM
MNP	Eyeball is yellow in color.	

Diseases of the Skin

See Infectious and Parasitic Diseases in Chapter 3, Section 2, Page 120.

Diseases of Muscles and Joints

MNT	Patient has pain in muscles of . . . (Table M-1 in Chapter 3, Section 4, Page 130).	
MNU	Patient has pain in joint(s) . . . (Table M-1 in Chapter 3, Section 4, Page 130).	
MNV	Patient has redness and swelling of joint(s) . . . (Table M-1 in Chapter 3, Section 4, Page 130).	
MNW	There is history of recent injury.	
MNX	There is no history of injury.	

Miscellaneous Illnesses

	Patient has had much alcohol	MLE
MOA	Patient is suffering from heat exhaustion.	
MOB	Patient is suffering from heat stroke.	
MOC	Patient is suffering from seasickness.	
MOD	Patient is suffering from exposure in lifeboat—indicate length of exposure (number) hours.	
MOE	Patient is suffering from frostbite . . . (Table M-1 in Chapter 3, Section 4, Page 130).	
MOF	Patient has been exposed to radioactive hazard.	

Code	*Meaning*	*Cross Reference*

Childbirth

MOK	I have a patient in childbirth aged . . . (number years).
MOL	Patient states she has had . . . (number) children.
MOM	Patient states child is due in . . . (number) weeks.
MON	Pains began . . . (number) hours ago.
MOO	Pains are feeble and produce no effect.
MOP	Pains are strong and effective.
MOQ	Pains are occurring every . . . (number) minutes.
MOR	The bag of membranes broke . . . (number) hours ago.
MOS	There is severe bleeding from the womb.
MOT	The head is coming first.
MOU	The buttocks are coming first.
MOV	A foot has appeared first.
MOW	An arm has appeared first.
MOX	The child has been born.
MOY	The child will not breathe.
MOZ	The placenta has been passed.
MPA	The placenta has not been passed.
MPB	I have a nonpregnant woman who is bleeding from the womb.

PROGRESS REPORT

MPE	I am carrying out prescribed instructions.
MPF	Patient is improving.
MPG	Patient is not improving.
MPH	Patient is relieved of pain.
MPI	Patient still has pain.
MPJ	Patient is restless.
MPK	Patient is calm.

CHAPTER 3.—MEDICAL SIGNAL CODE

Code	Meaning	Cross Reference
MPL	Symptoms have cleared.	
MPM	Symptoms have not cleared.	
MPN	Symptoms have increased.	
MPO	Symptoms have decreased.	
MPP	Treatment has been effective.	
MPQ	Treatment has been ineffective.	
MPR	Patient has died.	

CHAPTER 3

SECTION 3: MEDICAL ADVICE

REQUEST FOR ADDITIONAL INFORMATION

MQB	I cannot understand your signal; please use standard method of case description.
MQC	Please answer the following question(s).

DIAGNOSIS

MQE	My probable diagnosis is . . . (Table M-2 in Chapter 3, Section 4, Page 133).
MQF	My alternative diagnosis is . . . (Table M-2 in Chapter 3, Section 4, Page 133).
MQG	My probable diagnosis is infection or inflammation . . . (Table M-1 in Chapter 3, Section 4, Page 130).
MQH	My probable diagnosis is perforation of . . . (Table M-1 in Chapter 3, Section 4, Page 130).
MQI	My probable diagnosis is tumor of . . . (Table M-1 in Chapter 3, Section 4, Page 130).
MQJ	My probable diagnosis is obstruction of . . . (Table M-1 in Chapter 3, Section 4, Page 130).
MQK	My probable diagnosis is hemorrhage of . . . (Table M-1 in Chapter 3, Section 4, Page 130).
MQL	My probable diagnosis is foreign body in . . . (Table M-1 in Chapter 3, Section 4, Page 130).
MQM	My probable diagnosis is fracture of . . . (Table M-1 in Chapter 3, Section 4, Page 130).
MQN	My probable diagnosis is dislocation of . . . (Table M-1 in Chapter 3, Section 4, Page 130).
MQO	My probable diagnosis is sprain of . . . (Table M-1 in Chapter 3, Section 4, Page 130).
MQP	I cannot make a diagnosis.
MQT	Your diagnosis is probably right.
MQU	I am not sure about your diagnosis.

SPECIAL TREATMENT

MRI	You should refer to your International Ship's Medical Guide if available or its equivalent.
MRJ	You should follow treatment in your own medical guide.
MRK	You should follow the instructions for this procedure outlined in your own medical guide.
MRL	Commence artificial respiration immediately.
MRM	Pass catheter into bladder.

Code	Meaning	Cross Reference
MRN	Pass catheter again after . . . (number) hours.	
MRO	Pass catheter and retain it in bladder.	
MRP	Apply ice-cold compress and renew every . . . (number) hours.	
MRQ	Apply hot compress and renew every . . . (number) hours.	
MRR	Apply hot-water bottle to . . . (Table M-1 in Chapter 3, Section 4, Page 130).	
MRS	Insert ear drops . . . (number) times daily.	
MRT	Insert antiseptic eye drops . . . (number) times daily.	
MRU	Insert anesthetic eye drops . . . (number) times daily.	
MRV	Bathe eye frequently with hot water.	
MRW	Give frequent gargles one teaspoonful of salt in a tumblerful of water.	
MRX	Give enema.	
MRY	Do not give enema or laxative.	
MRZ	Was the result of the enema satisfactory?	
MSA	Give rectal saline slowly to replace fluid loss.	
MSB	Give subcutaneous saline to replace fluid loss.	
MSC	Apply well-padded splint(s) to immobilize limb. Watch circulation by inspection of color of fingers or toes.	
MSD	Apply cotton wool to armpit and bandage arm to side.	
MSF	Apply a sling and/or rest the part.	
MSG	Give light movements and massage daily.	
MSJ	Place patient in hot bath.	
MSK	To induce sleep give two sedative tablets.	
MSL	Reduce temperature of patient as indicated in general nursing chapter of Medical Guide.	
MSM	The swelling should be incised and drained.	
MSN	Dress wound with sterile gauze, cotton wool, and bandage.	
MSO	Dress wound with sterile gauze, cotton wool, and apply well-padded splint.	
MSP	Apply burn and wound dressing and bandage lightly.	
MSQ	Dress wound and bring edges together with adhesive plaster.	
MSR	The wound should be stitched.	

Code	*Meaning*	*Cross Reference*
MST	The wound should not be stitched.	
MSU	Stop bleeding by applying more cotton wool, firm bandaging, and elevation of the limb.	
MSV	Stop bleeding by manual pressure.	
MSW	Apply tourniquet for not more than fifteen minutes.	
MSX	Induce vomiting by giving an emetic.	
MSY	You should pass a stomach tube.	
MSZ	Do not try to empty stomach by any method.	

TREATMENT BY MEDICAMENTS

Prescribing

MTD	You should give . . . (Table M-3 in Chapter 3, Section 4, Page 134 and 135).	
MTE	You must not give . . . (Table M-3 in Chapter 3, Section 4, Page 134 and 135).	

Method of Administration and Dose

MTF	You should give one tablespoon (15 ml or $^1/_2$ oz.).	
MTG	You should give one dessertspoonful (7.5 ml or $^1/_4$ oz.).	
MTH	You should give one teaspoonful (4 ml or 1 drachm).	
MTI	You should give by mouth . . . (number) tablets/capsules.	
MTJ	You should give a tumblerful of water with each dose.	
MTK	You should give by intramuscular injection . . . (number) milligrams.	
MTL	You should give by subcutaneous injection . . . (number) milligrams.	
MTM	You should give by intramuscular injection . . . (number) ampoule(s).	
MTN	You should give by subcutaneous injection . . . (number) ampoule(s).	

Frequency of Dose

MTO	You should give once only.	
MTP	You should repeat after . . . (number) hours.	
MTQ	You should repeat every . . . (number) hours.	
MTR	You should continue for . . . (number) hours.	

Code	Meaning	Cross Reference

Frequency of External Application

MTT	You should apply once only.	
MTU	You should apply every . . . (number) hours.	
MTV	You should cease to apply.	
MTW	You should apply for . . . (number) minutes.	

DIET

MUA	Give nothing by mouth.	
MUB	Give water very freely.	
MUC	Give water only in small quantities.	
MUD	Give water only as much as possible without causing the patient to vomit.	
MUE	Give ice to suck.	
MUF	Give fluid diet, milk, fruit, juices, tea, mineral water.	
MUG	Give light diet such as vegetable soup, steamed fish, stewed fruit, milk puddings, or equivalent.	
MUH	Give normal diet as tolerated.	

CHILDBIRTH

MUI	Has she had previous children?	
MUJ	How many months pregnant is she?	
MUK	When did labor pains start?	
	Give enema.	MRX
MUL	Encourage her to rest between pains.	
MUM	Encourage her to strain down during pains.	
MUN	What is the frequency of pains (indicate in minutes).	
	To induce sleep give two sedative tablets.	MSK
MUO	Patient should strain down and you exert steady but gentle pressure on lower part of the abdomen but not on the womb to help expulsion of the placenta.	
MUP	You should apply tight wide binder around lower part of abdomen and hips.	
MUQ	You should apply artificial respiration gently by mouth technique on infant.	

Code	Meaning	Cross Reference

VACCINATION AGAINST SMALLPOX

MUR Has the patient been successfully vaccinated?

MUS Has the patient been vaccinated during the past three years?

MUT Patient has never been successfully vaccinated against smallpox.

MUU Patient was last vaccinated . . . (indicate date).

MUV Patient has vaccination marks.

GENERAL INSTRUCTIONS

MVA I consider the case serious and urgent.

MVB I do not consider the case serious or urgent.

MVC Put patient to bed lying down at absolute rest.

MVD Put patient to bed sitting up.

MVE Raise head of bed.

MVF Raise foot of bed.

MVG Keep patient warm.

MVH Keep patient cool.

MVI You should continue your local treatment.

MVJ You should continue your special treatment.

MVK You should continue giving . . . (Table M-3 in Chapter 3, Section 4, Pages 134 and 135).

MVL You should suspend your local treatment.

MVM You should suspend your special treatment.

MVN You should cease giving . . . (Table M-3 in Chapter 3, Section 4, Pages 134 and 135).

MVO You should isolate the patient and disinfect his cabin.

MVP You should land your patient at the earliest opportunity.

MVQ Patient should be seen by a doctor when next in port.

MVR I will arrange for hospital admission.

MVS I think I should come on board and examine the case.

MVT No treatment advised.

MVU Refer back to me in . . . (number) hours or before if patient worsens.

CHAPTER 3

SECTION 4: TABLES OF COMPLEMENTS

TABLE M-1—REGIONS OF THE BODY

Side of body or limb affected should be clearly indicated—right, left

FIGURE 1 (Front)

1. Frontal region of head	13. Arm upper	25. Scrotum
2. Side of head	14. Forearm	26. Testicles
3. Top of head	15. Wrist	27. Penis
4. Face	16. Palm of hand	28. Upper thigh
5. Jaw	17. Fingers	29. Middle thigh
6. Neck front	18. Thumb	30. Lower thigh
7. Shoulder	19. Central upper abdomen	31. Knee
8. Clavicle	20. Central lower abdomen	32. Patella
*9. Chest	*21. Upper abdomen	33. Front of leg
10. Chest, mid	*22. Lower abdomen	34. Ankle
11. Heart	*23. Lateral abdomen	35. Foot
12. Armpit	*24. Groin	36. Toes

FIGURE 2 (Back)

37. Back of head	44. Back of hand	51. Buttock
38. Back of neck	*45. Lower chest region	52. Anus
39. Back of shoulder	46. Spinal column upper	53. Back of thigh
40. Scapula region	47. Spinal column middle	54. Back of knee
41. Elbow	48. Spinal column lower	55. Calf
42. Back upper arm	*49. Lumbar (kidney) region	56. Heel
43. Back lower arm	50. Sacral region	

OTHER ORGANS OF THE BODY

57. Artery	69. Lip, lower	81. Tongue
58. Bladder	70. Lip, upper	82. Tonsils
59. Brain	71. Liver	83. Tooth, teeth
60. Breast	72. Lungs	84. Urethra
61. Ear(s)	73. Mouth	85. Uterus, womb
62. Eye(s)	74. Nose	86. Vein
63. Eyelid(s)	75. Pancreas	87. Voice box (larynx)
64. Gall bladder	76. Prostate	88. Whole abdomen
65. Gullet (esophagus)	77. Rib(s)	89. Whole arm
66. Gums	78. Spleen	90. Whole back
67. Intestine	79. Stomach	91. Whole chest
68. Kidney	80. Throat	92. Whole leg

*   Indicate side as required.

FIGURE 1

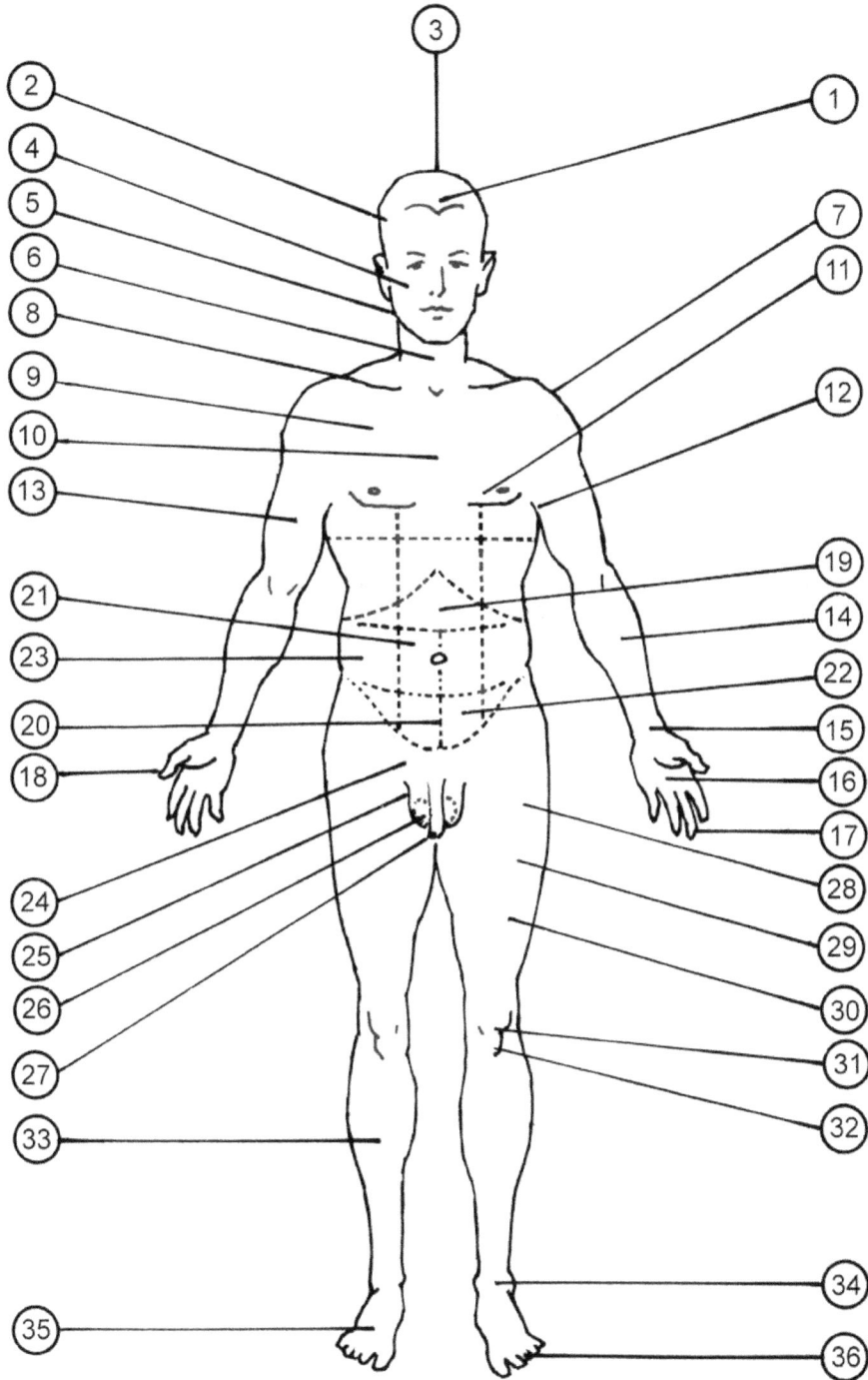

FIGURE 2

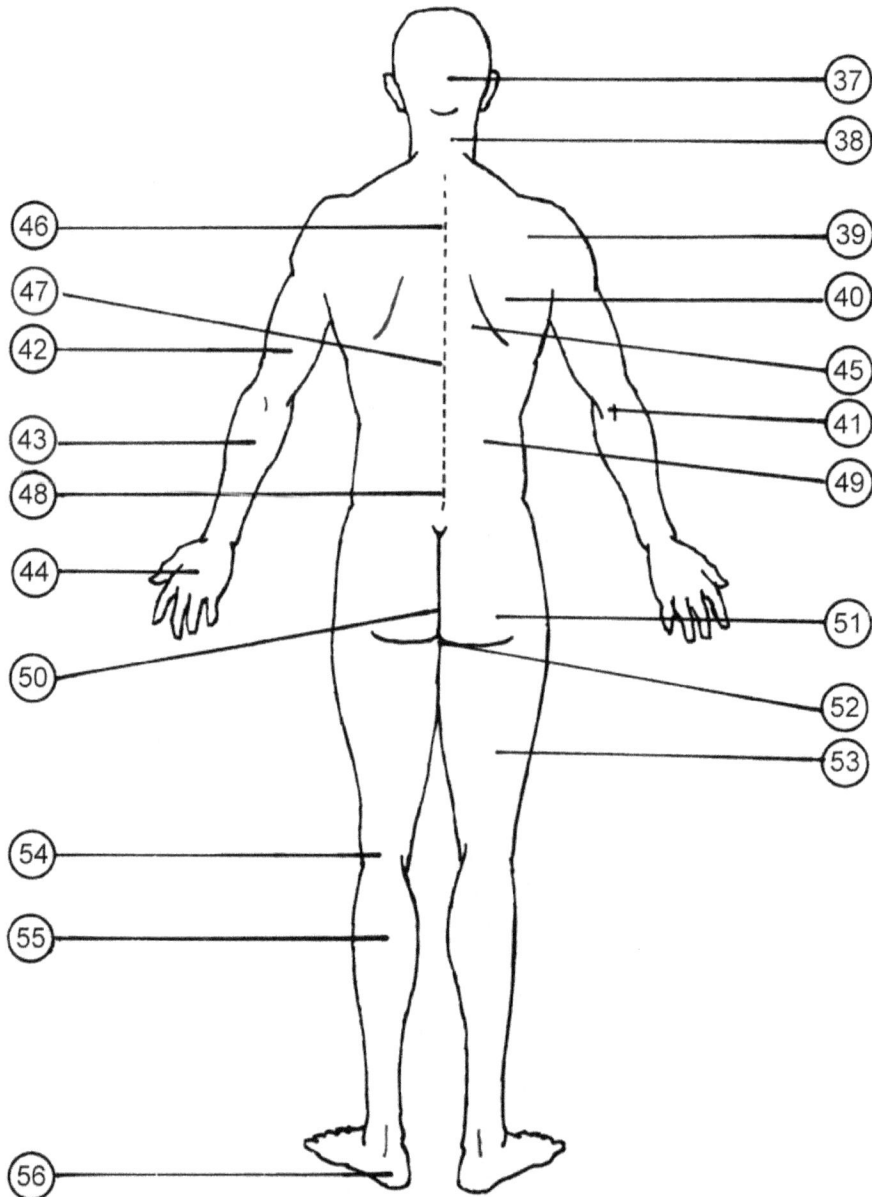

TABLE M-2—LIST OF COMMON DISEASES

1. Abscess
2. Alcoholism
3. Allergic reaction
4. Amoebic dysentery
5. Angina pectoris
6. Anthrax
7. Apoplexy (stroke)
8. Appendicitis
9. Asthma
10. Bacillary dysentery
11. Boils
12. Bronchitis (acute)
13. Bronchitis (chronic)
14. Brucellosis
15. Carbuncle
16. Cellulitis
17. Chancroid
18. Chicken pox
19. Cholera
20. Cirrhosis of the liver
21. Concussion
22. Compression of brain
23. Congestive heart failure
24. Constipation
25. Coronary thrombosis
26. Cystitis (bladder inflammation)
27. Dengue
28. Diabetes
29. Diabetic coma
30. Diptheria
31. Drug reaction
32. Duodenal ulcer
33. Eczema
34. Erysipelas
35. Fits
36. Gangrene
37. Gastric ulcer
38. Gastroenteritis
39. Gonorrhea
40. Gout
41. Heat cramps
42. Heat exhaustion
43. Heat stroke
44. Hepatitis
45. Hernia
46. Hernia (irreducible)
47. Hernia (strangulated)
48. Immersion foot
49. Impetigo
50. Insulin overdose
51. Indigestion
52. Influenza
53. Intestinal obstruction
54. Kidney stone (renal colic)
55. Laryngitis
56. Malaria
57. Measles
58. Meningitis
59. Mental illness
60. Migraine
61. Mumps
62. Orchitis
63. Peritonitis
64. Phlebitis
65. Piles
66. Plague
67. Pleurisy
68. Pneumonia
69. Poisoning (corrosive)
70. Poisoning (noncorrosive)
71. Poisoning (barbiturates)
72. Poisoning (methyl alcohol)
73. Poisoning (gases)
74. Poliomyelitis
75. Prolapsed intervertebral disc (slipped disc)
76. Pulmonary tuberculosis
77. Quinsy
78. Rheumatism
79. Rheumatic fever
80. Scarlet fever
81. Sciatica
82. Shingles (herpes zoster)
83. Sinusitis
84. Shock
85. Smallpox
86. Syphilis
87. Tetanus
88. Tonsillitis
89. Typhoid
90. Typhus
91. Urethritis
92. Urticaria (nettle rash)
93. Whooping cough
94. Yellow fever

TABLE M-3—LIST OF MEDICAMENTS*

FOR EXTERNAL USE

1. Auristillae Glyceris
   Glycerin ear drops
   EAR DROPS

2. Guttae Sulfacetamidi
   Sulfacetamide eye drops
   ANTISEPTIC EYE DROPS

3. Guttae Tetracainae
   Tetracaine eye drops
   ANAESTHETIC EYE DROPS

4. Linimentum Methylis Salicylatis
   Methyl salicylate liniment
   SALICYLATE LINIMENT

5. Lotio Calaminae
   Calamine Lotion
   CALAMINE LOTION

6. Lotio Cetrimidi
   Cetrimide lotion
   ANTISEPTIC LOTION

7. Naristillae Ephedrine
   Norephedrine hydrochloride drops
   NASAL DROPS

8. Paraffinum Molle Flavum
   Yellow soft paraffin
   SOFT PARAFFIN

9. Paraffinum Molle Flavum Carbasi Absorbentis
   Tulle gras dressing (Paraffin gauze)
   BURN/WOUND DRESSING

10. Unguentum Bacitracini
    Bacitracin ointment
    ANTIBIOTIC OINTMENT

11. Unguentum Benzocaini Compositum
    Compund benzocaine ointment
    PILE OINTMENT

12. Unguentum Xylocaini Hydrochloridi
    Mylocaine ointment
    LOCAL ANAESTHETIC OINTMENT

FOR INTERNAL USE

Allergic Conditions

13. Compressi Promethazini Hydrochloridi
    Promethazine hydrochloride tablets
    ANTIHISTAMINE TABLETS
    (25 mg per tablet)

14. Injectic Adrenalini
    Adrenaline injection
    ADRENALINE (1 mg in "Ampins")

CAUTION: THIS INJECTION NO. 14 TO BE USED ONLY ON MEDICAL ADVICE BY RADIO EXCEPT IN CASE OF ANAPHYLACTIC SHOCK DUE TO PENICILLIN INJECTION.

Antibiotics

15. Capsulae Tetracyclini Hydrochloridi
    Tetracycline hydrochloride capsules
    TETRACYCLINE CAPSULES
    (250 mg per capsule)

16. Compressi Phenoxymethylpenicillini
    Phenoxymethylpenicillin

PENICILLIN TABLETS
(125 mg per tablet)

17. Compressi Sulfadimidini
    Sulfadimidine tablets
    SULFONAMIDE TABLETS
    (500 mg per tablet)

18. Injectio Benzylpenicillini
    Procaine penicillin G
    PENICILLIN INJECTION
    (600,000 units per ampoule)

19. Injectio Streptomycini Sulfatis
    Streptomycin sulfate injection
    STREPTOMYCIN INJECTION
    (1,000 mg per ampoule)

20. Injectio Tetracyclini hydrochloridi
    Tetracycline hydrochloride
    TETRACYCLINE INJECTION
    (100 mg per ampoule)

---

* Preparations listed may have been substituted by equivalent preparations in the ship's medicine chest. For the sake of uniformity, medicaments are indicated in the first place by their Latin denominations so that a correct translation can be found in each language.

SECTION 4.—TABLES OF COMPLEMENTS

Asthma

21. Compressi Aminophyllini
Aminophylline tablets
ASTHMA RELIEF TABLETS
(300 mg per tablet)

CAUTION: THIS TABLET NO. 21 TO BE USED ONLY
ON MEDICAL ADVICE BY RADIO.

22. Compressi Ephedrini Hydrochloridi
Ephedrine Hydrochloride tablets
EPHEDRINE TABLETS
(30 mg per tablet)

23. Tinctura Benzoini Composita
Tincture of benzoin compound
INHALATION MIXTURE

Cough

24. Compressi Codeini Phosphatis
Codein phosphate tablets
CODEIN TABLETS
(15 mg per tablet)

25. Linctus Scillae Opiata
Linctus of squill, opiate
COUGH LINCTUS

Diarrhea

26. Mistura Kaolini et Morphinae
Kaolin and morphine mixture
DIARRHEA MIXTURE

Heart

27. Compressi Glycerylis Trinitratis
Glycerin Trinitrate tablets
HEART TABLETS
(0.5 mg per tablet)

*NOTE: For congestive heart failure the following prepara-
tions are available on board ship, but they should be
used only on medical advice transmitted in plain language and not by Code:*

Compressi Chlorothiazidi (Chlorothiazide) or equivalent (500 mg per tablet)
Compressi Digoxin (Digoxin tablets) or equivalent
(0.25 mg per tablet)

Indigestion

28. Compressi Magnesii Trisilicas
Magnesium trisilicate
STOMACH TABLETS

Laxatives

29. Compressi Colocynthidis et Jalapae Compositae
Counpound Colocynth and Jalap tablets
VEGETABLE LAXATIVE TABLETS

30. Magnesii Hydroxidum
Magnesium hydroxide mixture
LIQUID LAXATIVE—"Milk of Magnesia"

Malaria

31. Compressi Chloroquini Sulfatis
Chloroquine sulfate tablets
MALARIA TABLETS
(200 mg per tablet)

Pain

32. Compressi Acidi Acetylasalicylici
Acetylsalicylic acid tablets
ASPIRIN TABLETS
(300 mg per tablet)

33. Injectio Morphini
Morphine sulfate injection
MORPHINE INJECTION
(15 mg per ampoule)

Sedation

34. Compressi Butobarbitali
Butobarbitone tablets
SEDATIVE TABLETS
(100 mg per tablet)

35. Compressi Phenobarbitali
Phenobarbitone tablets
PHENOBARBITONE TABLETS
(30 mg per tablet)

36. Compressi Chlorpromazini Hydrochloridi
Chlorpromazine hydrochloride tablets
TRANQUILLIZER TABLETS (LARGACTIL)
(50 mg per tablet)

CAUTION: THIS TABLET NO. 36 TO BE USED ONLY
ON MEDICAL ADVICE BY RADIO.

Salt Depletion or Heat Cramps

37. Compressi Natrii Chloridi Solv
Sodium chloride tablets
SALT TABLETS
(500 mg per tablet)

Seasickness

38. Compressi Hyoscini Hydrobromidi
Hysocine Hydrobromide tablets
SEASICKNESS TABLETS
(0.3 mg per tablet)

CHAPTER **4**

CHAPTER 4
DISTRESS AND LIFESAVING SIGNALS AND
RADIOTELEPHONE PROCEDURES

CHAPTER 4

SECTION 1: DISTRESS SIGNALS

(PRESCRIBED BY THE INTERNATIONAL REGULATIONS FOR PREVENTING COLLISIONS AT SEA 1972)

To be used or displayed, either together or separately, by a vessel (or seaplane on the water) in distress requiring assistance from other vessels or from the shore.

1. A gun or other explosive signal fire at intervals of about a minute.
2. A continuous sounding with any fog-signaling apparatus.
3. Rockets or shells, throwing red stars fired one at a time at short intervals.
4. A signal made by radiotelegraphy or by any other signaling method consisting of the group • • • – – – • • • **SOS** in the Morse Code.
5. A signal sent by radiotelephony consisting of the spoken word **"MAYDAY"**.
6. The International Code Signal of distress indicated by **NC**.
7. A signal consisting of a square flag having above or below it a ball or anything resembling a ball.
8. Flames on the vessel (as from a burning tar barrel, oil barrel, etc.).
9. A rocket parachute flare or a hand flare showing a red light.
10. A smoke signal giving off a volume of orange-colored smoke.
11. Slowly and repeatedly raising and lowering arms outstretched to each side.
12. The radiotelegraph alarm signal.[*]
13. The radiotelephone alarm signal.[**]
14. Signals transmitted by emergency position-indicating beacons.[***]

NOTES: (a) Vessels in distress may use the radiotelegraph alarm signal or the radiotelephone alarm signal to secure attention to distress calls and messages. The radiotelegraph alarm signal, which is designed to actuate the radiotelegraph auto alarms of vessels so fitted, consists of a series of twelve dashes, sent in 1 minute, the duration of each dash being 4 seconds and the duration of the interval between 2 consecutive dashes being 1 second. The radiotelephone alarm signal consists of 2 tones transmitted alternately over periods of from 30 seconds to 1 minute.

*(b)*The use of any of the foregoing signals, except for the purpose of indicating that a vessel or seaplane is in distress, and the use of any signals which may be confused with any of the above signals is prohibited.

*(c)*Attention is drawn to the relevant sections of the Merchant Ship Search and Rescue Manual and the following signals:
   (i.) a piece of orange-colored canvas with either a black square and circle or other appropriate symbol (for identification from the air);
   (ii.) a dye marker.

---

[*] A series of twelve four second dashes at intervals of one second.

[**] Two audio tones transmitted alternately at frequency of 2200 Hz and 1300 Hz for a duration of 30 seconds to one minute.

[***] Either the signal described in [**] above or a series of single tones at a frequency of 1300 Hz.

CHAPTER 4

SECTION 2: TABLE OF LIFESAVING SIGNALS

I LANDING SIGNALS FOR THE GUIDANCE OF SMALL BOATS WITH CREWS OR PERSONS IN DISTRESS

	MANUAL SIGNALS	LIGHT SIGNALS	OTHER SIGNALS	SIGNIFICATION
Day Signals	Vertical motion of a white flag or of the arms	or firing of a **green** star signal	or code letter **K** given by light or sound-signal apparatus	This is the best place to land
Night Signals	**Vertical** motion of a white light or flare	or firing of a **green** star signal	or code letter **K** given by light or sound-signal apparatus	

A range (indication of direction) may be given by placing a steady white light or flare at a lower level and in line with the observer

	MANUAL SIGNALS	LIGHT SIGNALS	OTHER SIGNALS	SIGNIFICATION
Day Signals	**Horizontal** motion of a white flag or of the arms extended horizontally	or firing of a **red** star signal	or code letter **S** given by light or sound-signal apparatus	Landing here highly dangerous
Night Signals	**Horizontal** motion of a light or flare	or firing of a **red** star signal	or code letter **S** given by light or sound-signal apparatus	
Day Signals	**1 Horizontal** motion of a white flag, followed by **2** the placing of the white flag in the ground and **3** by the carrying of another white flag in the direction to be indicated	**1** or firing of a red star signal vertically and **2** a white star signal in the direction towards the better landing place	**1** or signalling the code letter **S** (...) followed by the code letter **R** (. _ .) if a better landing place for the craft in distress is located more to the *right* in the direction of approach **2** or signaling the code letter **S** (...) followed by the code letter **L** (. _ ..) if a better landing place for the craft in distress is located more to the *left* in the direction of approach	Landing here highly dangerous. A more favorable location for landing is in the direction indicated
Night Signals	**1 Horizontal** motion of a white light or flare **2** followed by the placing of the white light or flare on the ground and **3** the carrying of another white light or flare in the direction to be indicated	**1** or firing of a **red** star signal vertically and a **2 white** star signal in the direction towards the better landing place	**1** or signalling the code letter **S** (...) followed by the code letter **R** (. _ .) if a better landing place for the craft in distress is located more to the *right* in the direction of approach **2** or signaling the code letter **S** (...) followed by the code letter **L** (. _ ..) if a better landing place for the craft in distress is located more to the *left* in the direction of approach	

II SIGNALS TO BE EMPLOYED IN CONNECTION WITH THE USE OF SHORE LIFESAVING APPARATUS

	MANUAL SIGNALS	LIGHT SIGNALS	OTHER SIGNALS	SIGNIFICATION
Day Signals	Vertical motion of a white flag or of the arms	or firing of a green star signal		In general: affirmative Specifically: rocket line is held — tail block is made fast — hawser is made fast — man is in the breeches buoy — haul away
Night Signals	Vertical motion of a white light or flare	or firing of a green star signal		
Day Signals	Horizontal motion of a white flag or of the arms extended horizontally	or firing of a red star signal		In general: negative Specifically: slack away - avast hauling
Night Signals	Horizontal motion of a white light or flare	or firing of a red star signal		

III REPLIES FROM LIFESAVING STATIONS OR MARITIME RESCUE UNITS TO DISTRESS SIGNALS MADE BY A SHIP OR PERSON

Day Signals		Orange smoke signal	or combined *light* and *sound* signal (thunder-light) consisting of 3 single signals which are fired at intervals of approximately one minute	You are seen - assistance will be given as soon as possible (Repetition of such signal shall have the same meaning)
Night Signals		White star rocket consisting of 3 single signals which are fired at intervals of approximately one minute		

If necessary, the day signals may be given at night or the night signals by day

IV AIR-TO-SURFACE VISUAL SIGNALS

Signals used by aircraft engaged in search and rescue operations to direct ships towards an aircraft, ship or person in distress

PROCEDURES PERFORMED IN SEQUENCE BY AN AIRCRAFT			SIGNIFICATION
1 CIRCLE the vessel at least once.	2 CROSS the vessel's projected course close AHEAD at a low altitude while ROCKING the wings. (See Note)	3 HEAD in the direction in which the vessel is to be directed.	The aircraft is directing a vessel towards an aircraft or vessel in distress (Repetition of such signals shall have the same meaning)
4 CROSS the vessel's wake close ASTERN at low altitude while ROCKING the wings. (See Note)			The assistance of the vessel is no longer required (Repetition of such signals shall have the same meaning)
NOTE - Opening and closing the throttle or changing the propeller pitch may also be practiced as an alternative means of attracting attention to that of rocking the wings. However, this form of sound signal may be less effective than the visual signal of rocking the wings owing to high noise level on board the vessel.			

Signals used by a vessel in response to an aircraft engaged in search and rescue operations · **SIGNIFICATION**

Hoist "Code and Answering" pendant Close up; or	Change the heading to the required direction; or	Flash Morse Code signal "T" by signal lamp.	Acknowledges receipt of aircraft's signal
Hoist international flag "N" (NOVEMBER); or		Flash Morse Code signal "N" by signal lamp.	Indicates inability to comply

V SURFACE-TO-AIR VISUAL SIGNALS

Communication from surface craft or survivors to an aircraft

Use International Code of Signals or plain language by use of a torch, signalling lamps or signal flags.	or	Use the following surface-to-air visual signals by displaying the appropriate signal on the deck or on the ground.	
Message		International Code of Signals	ICAO* visual symbols
- Require assistance		V	V
- Require medical assistance		W	X
- No or negative		N	N
- Yes or affirmative		C	Y
- Proceeding in this direction			↑

* ICAO annex 12 — Search and rescue

Reply from an aircraft observing the above signals from surface craft or survivors

SIGNIFICATION

Drop a message or	Rock the wings (during daylight) or	Flash the landing lights or navigation lights on and off twice (during hours of darkness) or	Flash Morse Code signal "T" or "R" by light or	Use any other suitable signal	**Message understood**
Fly straight and level without rocking wings or	Flash Morse Code Signal "RPT" by light or	Use any other suitable signal			**Message not understood (repeat)**

VI SIGNALS TO SURVIVORS

Procedures performed by an aircraft   **SIGNIFICATION**

Drop a message or	Drop communication equipment suitable for establishing direct contact		**The aircraft wishes to inform or instruct survivors**

* High visibility colored streamer

Signals used by survivors in response to a message dropped by an aircraft   **SIGNIFICATION**

Flash Morse Code signal "T" or "R" by light or	Use any other suitable signal		**Dropped messages is understood by the survivors**
Flash Morse Code signal "RPT" by light			**Dropped messages is not understood by the survivors**

144

"CONFLICT AND PERMANENT IDENTIFICATION OF RESCUE CRAFT" [*]

Shape, color, and positioning of emblem for medical transports

1. The following emblems can be used separately or together to show that a vessel is protected as a medical transport under the Geneva Convention:

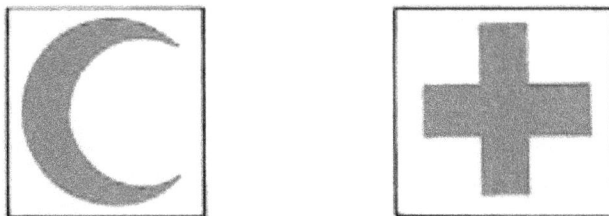

2. The emblem, positioned on the vessel's sides, bow, stern, and deck, shall be painted dark red on a white background.
   a.   On the vessel's sides the emblem shall extend from the waterline to the top of the ship's hull.
   b.   The emblems on the vessel's bow and stern must, if necessary, be painted on a wooden structure so as to be clearly visible to other vessels ahead or astern.
   c.   The deck emblem must be as clear of the vessel's equipment as possible to be clearly visible from aircraft.
3. In order to provide the desired contrast for infrared film or instruments, the red emblem must be painted on top of a black primer paint.
4. Emblems may also be made of materials which make them recognizable by technical means of detecting.

Illumination

1. At night and in restricted visibility the emblems shall be illuminated or lit.
2. At night and in restricted visibility all deck and overside lights must be fully lit to indicate that the vessel is engaged in medical operations.

Personal equipment

1. Subject to the instructions of the competent authority, medical and religious personnel carrying out their duties in the battle area shall, as far as possible, wear headgear and clothing bearing the distinctive emblem.

Flashing blue light for medical transports

1. A vessel engaged in medical operations shall exhibit one or more all-around flashing blue lights of the color prescribed in paragraph 4.
2. The visibility of the lights shall be as high as possible and not less than 3 nautical miles in accordance with Annex 1 to the International Regulations for Preventing Collisions at Sea, 1972.
3. The light or lights shall be exhibited as high above the hull as practical and in such a way that at least one light shall be visible from any direction.
4. The recommended blue color is obtained by using, as trichromatic co-ordinates:

green boundary	$y = 0.065+0.805x$
white boundary	$y = 0.400-x$
purple boundary	$x = 0.133+0.600y$

5. The frequency of the flashing light shall be between 60 and 100 flashes per minute.

Radar transponders

1. It should be possible for medical transports to be identified by other vessels equipped with radar by signals from a radar tran-

---

[*]   In accordance with Article 27 of the Second Geneva Convention of 12 August 1949, this chapter also applies to coastal rescue craft.

sponder fitted on the medical transport.

2. The signal from the medical transport transponder shall consist of the group YYY, in accordance with article 40 of the Radio Regulations, followed by the call sign or other recognized means of identification.

Underwater acoustic signals

1. It should be possible for medical transports to be identified by submarines by appropriate underwater signals transmitted by the medical transports.

2. The underwater signal shall consist of the call sign of the ship preceded by the single group YYY transmitted in Morse on an appropriate acoustic frequency, e.g., 5 kHz.

Rescue craft carried by medical transports

1. Every rescue craft should be equipped with a mast on which a Red Cross flag measuring about 2 x 2 meters can be hoisted.

Flashing blue light for medical aircraft

1. The light signal, consisting of a flashing blue light, is established for the use of medical aircraft to signal their identity. No other aircraft shall use this signal. The recommended flashing rate of the blue light is between sixty and one hundred flashes per minute.

2. Medical aircraft should be equipped with such lights as may be necessary to make the light signal visible in as many directions as possible.

CHAPTER 4

SECTION 3: RADIOTELEPHONE PROCEDURES

RECEPTION OF SAFETY MESSAGES

Any message which you hear prefixed by one of the following words concerns SAFETY:

MAYDAY (Distress)	Indicates that a ship, aircraft, or other vehicle is threatened by grave and imminent danger and requests immediate assistance.
PAN (Urgency)	Indicates that the calling station has a very urgent message to transmit concerning the safety of a ship, aircraft, or other vehicle, or the safety of a person.
SECURITE (Safety)	Indicates that the station is about to transmit a message concerning the safety of navigation or giving important meteorological warnings.

If you hear these words, pay particular attention to the message and call the master or the officer on watch.

DISTRESS TRANSMITTING PROCEDURES

To be used only if IMMEDIATE ASSISTANCE is required:

USE PLAIN LANGUAGE WHENEVER POSSIBLE. If language difficulties are likely to arise use Tables 2 and 3 on Page 149, sending the word INTERCO to indicate that the message will be in the International Code of Signals.
Call out letters as in Table 1 on Page 148. Call out numbers figure by figure as in Table 1.

To indicate DISTRESS:

1. If possible transmit the ALARM SIGNAL (i.e., two-tone signal) for 30 seconds to one minute, but do not delay the message if there is insufficient time in which to transmit the Alarm Signal.
2. Send the following DISTRESS CALL:
Mayday Mayday Mayday. This is . . . (name or call sign of ship spoken three times).
3. Then send the DISTRESS MESSAGE composed of:
Mayday followed by the name or call sign of ship;
Position of ship;
Nature of distress;
And, if necessary, transmit the nature of the aid required and any other information which will help the rescue.

EXAMPLES OF DISTRESS PROCEDURE

1. Where possible, transmit ALARM SIGNAL followed by spoken words Mayday Mayday Mayday. This is . . . (name of ship spoken three times, or call sign of ship spelled three times using TABLE 1, on Page 148) Mayday . . . (name or call sign of ship) Position 54 25 North 016 33 West I am on fire and require immediate assistance.

2. Where possible, transmit ALARM SIGNAL followed by spoken words Mayday Mayday Mayday . . . (name of ship spoken three times, or call sign of ship spelled three times using TABLE 1) Mayday . . . (name or call sign of ship) Interco Alfa Nadazero Unaone Pantafive Ushant Romeo Kartefour Nadazero Delta X-ray. "(Ship) in Distress Position 015 Degrees Ushant 40 miles I am sinking."

3. Where possible, transmit ALARM SIGNAL followed by spoken words Mayday Mayday Mayday . . . (name of ship spoken three times, or call sign of ship spelled three times using TABLE 1) Mayday . . . (name or call sign of ship) Interco Lima Pantafive Kartefour Bissotwo Pantafive November Golf Nadazero Unaone Soxisix Terrathree Terrathree Whiskey Charlie Bravo Soxisix. "(Ship) in Distress Position Latitude 54 25 North Longitude 016 33 West I require immediate assistance I am on fire."

TABLE 1
PHONETIC ALPHABET AND FIGURE SPELLING TABLES
(May be used when transmitting plain language or code.)

Letter	Word	Pronounced as	Letter	Word	Pronounced as
A	Alfa	**AL** FAH	N	November	NO **VEM** BER
B	Bravo	**BRAH** VOH	O	Oscar	**OSS** CAH
C	Charlie	**CHAR** LEE or	P	Papa	PAH **PAH**
		SHAR LEE	Q	Quebec	KEH **BECK**
D	Delta	**DELL** TAH	R	Romeo	**ROW** ME OH
E	Echo	**ECK** OH	S	Sierra	SEE **AIR** RAH
F	Foxtrot	**FOKS** TROT	T	Tango	**TANG** GO
G	Golf	GOLF	U	Uniform	**YOU** NEE FORM or
H	Hotel	HOH **TELL**			**OO** NEE FORM
I	India	**IN** DEE AH	V	Victor	**VIK** TAH
J	Juliett	**JEW** LEE **ETT**	W	Whiskey	**WISS** KEY
K	Kilo	**KEY** LOH	X	X-ray	**ECKS** RAY
L	Lima	**LEE** MAH	Y	Yankee	**YANG** KEY
M	Mike	MIKE	Z	Zulu	**ZOO** LOO

NOTE: The syllables to be emphasized are **boldfaced**.

Figure or Mark to be Transmitted	Word	Pronounced as	Figure or Mark to be Transmitted	Word	Pronounced as
0	NADAZERO	NAH-DAH-ZAY-ROH	6	SOXISIX	SOK-SEE-SIX
1	UNAONE	OO-NAH-WUN	7	SETTESEVEN	SAY-TAY-SEVEN
2	BISSOTWO	BEES-SOH-TOO	8	OKTOEIGHT	OK-TOH-AIT
3	TERRATHREE	TAY-RAH-TREE	9	NOVENINE	NO-VAY-NINER
4	KARTEFOUR	KAR-TAY-FOWER	**Decimal point**	DECIMAL	DAY-SEE-MAL
5	PANTAFIVE	PAN-TAH-FIVE	**Full stop**	STOP	STOP

NOTE: Each syllable should be equally emphasized.

SECTION 3.—RADIOTELEPHONE PROCEDURES

TABLE 2

Position in Code

(1) **By Bearing and Distance from a Landmark**

Code letter **A** (Alfa) followed by a three-figure group for ship's TRUE bearing from landmark;

Name of landmark:

Code letter **R** (Romeo) followed by one or more figures for distance in nautical miles.

or

(2) **By Latitude and Longitude**

Latitude

Code letter **L** (Lima) followed by a four-figure group; (2 figures for Degrees, 2 figures for Minutes) and either—**N** (November) for Latitude North, or **S** (Sierra) for Latitude South.

Longitude

Code letter **G** (Golf) followed by a five-figure group; (3 figures for Degrees, 2 figures for Minutes) and either—**E** (Echo) for Longitude East, or **W** (Whiskey) for Longitude West.

TABLE 3

Nature of Distress in Code

Code Letters	Words to be transmitted	Text of Signal
AE	Alfa Echo	I must abandon my vessel.
BF	Bravo Foxtrot	Aircraft is ditched in position indicated and requires immediate assistance.
CB	Charlie Bravo	I require immediate assistance.
CB6	Charlie Bravo Soxisix	I require immediate assistance, I am on fire.
DX	Delta X-ray	I am sinking.
HW	Hotel Whiskey	I have collided with surface craft.
Answer to Ship in Distress		
CP	Charlie Papa	I am proceeding to your assistance.
ED	Echo Delta	Your distress signals are understood.
EL	Echo Lima	Repeat the distress position.

NOTE: A more comprehensive list of signals may be found in Chapter 2.

APPENDIX
US/RUSSIA SUPPLEMENTARY SIGNALS FOR NAVAL VESSELS

IR	1	I am engaged in oceanographic operations.
IR	2	I am streaming/towing hydrographic survey equipment . . . . meters astern.
IR	3	I am recovering hydrographic survey equipment.
IR	4	I am conducting salvage operations.
JH	1	I am attempting to retract a grounded vessel.
MH	1	Request you not to cross my course ahead of me.
NB	1	I have my unattached hydrographic survey equipment bearing in a direction from me as indicated. . . . (Table 3 of ICS).
PJ	1	I am unable to alter course to my starboard.
PJ	2	I am unable to alter course to my port.
PJ	3	Caution, I have a steering casualty.
PP	8	Dangerous operations in progress. Request you remain clear of the hazard which is in the direction from me as indicated. . . . (Table 3 of ICS).
QF	1	Caution, I have stopped engines.
QS	6	I am proceeding to anchorage on course. . . .
QV	2	I am in a fixed multiple leg moor using two or more anchors or buoys fore and aft. Request you remain clear.
QV	3	I am anchored in deep water with hydrographic survey equipment streamed.
RT	2	I intend to pass you on your port side.
RT	3	I intend to pass you on your starboard side.
RT	4	I will overtake you on your port side.
RT	5	I will overtake you on your starboard side.
RT	6	I am/Formation is maneuvering. Request you remain clear of the hazard which is in the direction from me as indicated. . . . (Table 3 of ICS).
RT	7	I shall approach your ship on starboard side to a distance of. . . . 100's of meters (yards).
RT	8	I shall approach your ship on port side to a distance of. . . . 100's of meters (yards).
RT	9	I shall cross astern at a distance of. . . . 100's of meters (yards).
RU	2	I am beginning a port turn in approximately. . . . minutes.
RU	3	I am beginning a starboard turn in approximately. . . . minutes.
RU	4	The formation is preparing to alter course to port.
RU	5	The formation is preparing to alter course to starboard.
RU	6	I am engaged in maneuvering exercises. It is dangerous to be inside the formation.
RU	7	I am preparing to submerge.
RU	8	A submarine will surface within two miles of me within 30 minutes. Request you remain clear.
TX	1	I am engaged in fisheries patrol.
SL	2	Request your course, speed, and passing intention.
UY	1	I am preparing to launch/recover aircraft on course. . . .
UY	2	I am preparing to conduct missile exercises. Request you remain clear of the hazard which is in the direction from me as indicated. . . . (Table 3 of ICS).
UY	3	I am preparing to conduct gunnery exercises. Request you remain clear of the hazard which is in the direction from me as indicated. . . . (Table 3 of ICS).
UY	4	I am preparing to conduct/am conducting operations employing explosive charges.
UY	5	I am maneuvering in preparation for torpedo launching exercises. Request you remain clear of the hazard which is in the direction from me as indicated . . . . (Table 3 of ICS).
UY	6	I am preparing to conduct/am conducting underway replenishment on course. . . . Request you remain clear.
UY	7	I am preparing to conduct extensive small boat and ship to shore amphibious training operations.
UY	8	I am maneuvering to launch/recover landing craft/boats.
UY	9	I am preparing to conduct/am conducting helicopter operations over my stern.
UY	10	I am testing my gun systems.
UY	11	I am testing my missile systems.
UY	12	I am preparing to conduct/am conducting gunnery/bombing exercises from aircraft on a towed target. Request you remain clear of the hazard which is in the direction from me as indicated . . . (Table 3 of ICS).
ZL	1	I have received and understood you message.
ZL	2	Do you understand? Request acknowledgment.

Special Warning Signals.—The following signals are used by Russian naval vessels to warn foreign vessels that they have violated **Regulations for entry, navigating and stopping in Russian Territorial Waters (Territorial Sea) or Russian Inland Waters:**

SNG	You have violated the Russian borders. I demand that you leave Russian waters immediately.
SNO	I demand that you leave Russian waters immediately. Unless you do so, a force of arms will be used against you.
SNP	You are violating the regulations for navigating and remaining in Russian waters. I demand that you cease violations.
SNR	Despite warnings, you continue to violate the regulations for navigating and remaining in Russian waters. You are to leave them immediately.

By day these signals will be made by flags of the International Code of Signals. By night they will be in Morse Code by signal lamp. In addition, the signals may be transmitted by RT in plain language on 500 kHz, 2182 kHz and 156.8 MHz, as well as by voice using megaphone or any other amplifying device.

Warning signals to foreign submarines which are submerged:

The signal of two series of explosions with three explosions in each series (where the interval between the explosions in a series is one minute and the interval between the series is three minutes), means: You are in Russian waters. I demand that you surface immediately. Unless you comply with this order within 10 minutes, a force of arms will be used against you.

An acoustic signal by sonar may be given simultaneously, with the same meaning as described above. The signal will consist of five dashes, each dash three seconds long, interval between dashes, three seconds.

INDEXES

155

INDEX
MEDICAL SIGNAL CODE

(Numbers Refer to Pages)

www.ingramcontent.com/pod-product-compliance
Lightning Source LLC
Chambersburg PA
CBHW051217200326
41519CB00025B/7143